IoT POTENTIAL FOR GREEN ENERGY SOLUTIONS

IoT POTENTIAL FOR GREEN ENERGY SOLUTIONS

Abhishek Kumar, PhD

Ajay Roy, PhD

Shamimul Qamar, PhD

Editors

AAP | APPLE ACADEMIC PRESS

First edition published 2026

Apple Academic Press Inc.
1265 Goldenrod Circle, NE,
Palm Bay, FL 32905 USA

760 Laurentian Drive, Unit 19,
Burlington, ON L7N 0A4, Canada

CRC Press
2385 NW Executive Center Drive,
Suite 320, Boca Raton FL 33431

4 Park Square, Milton Park,
Abingdon, Oxon, OX14 4RN, UK

Library and Archives Canada Cataloguing in Publication

Title: IoT potential for green energy solutions / Abhishek Kumar, PhD, Ajay Roy, PhD, Shamimul Qamar, PhD, editors.
Names: Kumar, Abhishek (Associate Professor), editor. | Roy, Ajay, editor. | Qamar, Shamimul, editor.
Description: First edition. | Includes bibliographical references and index.
Identifiers: Canadiana (print) 20250206676 | Canadiana (ebook) 20250206706 | ISBN 9781779643582 (hardcover) | ISBN 9781779643599 (ebook)
Subjects: LCSH: Internet of things—Environmental aspects. | LCSH: Internet of things—Industrial applications. | LCSH: Clean energy.
Classification: LCC TK5105.8857 .I62 2026 | DDC 004.67/8—dc23

Library of Congress Cataloging-in-Publication Data

CIP data on file with US Library of Congress

ISBN: 978-1-77964-358-2 (hbk)
ISBN: 978-1-77964-359-9 (ebk)
DOI: 10.1201/9781779643599

About the Editors

Abhishek Kumar, PhD
Associate Professor, School of Electronics and Electrical Engineering, Lovely Professional University, Phagwara, Punjab, India

Abhishek Kumar, PhD, is an Associate Professor in the School of Electronics and Electrical Engineering at Lovely Professional University, India, from which he earned a PhD in VLSI Design for Low Power and Secured Architecture. His academic journey includes earning an MTech in Electronics Engineering from the University of Mumbai, India, and graduating from The Institution of Engineers. With over 10 years of teaching experience in VLSI design, Dr. Kumar has contributed significantly to the field. He has authored more than 60 research papers published in refereed journals and conferences, with four books included in the Scopus database. Notably, Dr. Kumar has secured five copyrights and published 25 patents. He is actively engaged in organizing workshops, summer internships, and expert lectures for students. Dr. Kumar has also served as a session chair, conference steering committee member, editorial board member, and reviewer for international/national IEEE journals and conferences. His expertise spans various domains including CMOS VLSI design, low power VLSI circuit design, machine learning, VLS, cryptanalysis, physical design, IoT, embedded system design, memory design, data converters, and ASIC-SoC-FPGA.

Ajay Roy, PhD
Professor, Department of Electrical Engineering, College of Engineering, University of Bisha, P.O. Box 551, Bisha, Kingdom of Saudi Arabia

Ajay Roy, PhD, received an engineering degree from the West Bengal University of Technology and a PhD degree in electronics from Dr. B.R. Ambedkar National

Institute of Technology, Jalandhar, India. He is currently associated with University of Bisha, Bisha, Kingdom of Saudi Arabia as a Professor with more than 14 years of experience in academics. His areas of expertise include embedded systems, robotics, wireless sensor networks, the Internet of Things, and machine learning. He has published more than 100 research papers in refereed journals/conferences. He has been honored as a keynote speaker and the session chair of international/national conferences, faculty development programs, workshops, and webinars. He has guided four PhD scholars and is currently guiding nine PhD scholars.

Shamimul Qamar, PhD

Professor, Faculty of Computer Science and Engineering, College of Science and Arts, Dhahran Al Janub, King Khalid University, Abha, Saudi Arabia; Consultant, Jackson State University, Missississippi, USA

Shamimul Qamar, PhD, is a Professor and an eminent scholar in the field of Computer Science and Engineering. He holds a BTech from MMMTU Gorakhpur, MTech from Aligarh Muslim University, Aligarh, and his PhD in Computer Science and Engineering degree from IIT Roorkee, India, with a highly honorable grade. Prof. Qamar has a wide teaching experience at various engineering colleges. He has research interests in computer networks and multimedia, networking and security, and artificial intelligence. He has published several research papers in reputed national/international journals and conferences. He served as a consultant at Jackson State University, USA. He is a reviewer for journals published by IEEE, Elsevier, MDPI, EUROSIP, as well as for the *International Journal of Computer Science and Information Security* (IJCSIS). He has written several textbooks and chapters in the field of electronics and computer engineering. He is also a technical program committee member for various international conferences. He is a lifetime member of the International Association of Engineers and a life member of the Indian Society of Technical Education. His technical depth and interest resulted in setting up a research lab according to the latest technical innovations. Along with this, he has actively participated in various technical courses workshops, seminars, etc. at the IITS.

Contents

Contributors

Manoj Singh Adhikari
School of Electronics and Electrical Engineering, Lovely Professional University, Phagwara, Punjab, India

Valli Bhasha
Department of Electronics and Communication Engineering, Sri Venkateswara College of Engineering, Andhra Pradesh, India

Naman N. Bhavsar
Civil Engineering Department, GTU-ITR, Mehsana, Gujarat, India

Biswaranjan Bhola
Department of Computer Science and Engineering, Gandhi Institute of Engineering and Technology University, Gunupur, Odisha, India

Reena Chandel
Rayat Bahra University, Punjab, India

Pratpsinh Chougule
Department of Automation & Robotics, Sharad Institute of Technology, College of Engineering, Yadrav, Ichalkaranji, Kolhapur, Maharashtra, India

Ashish A. Desai
Mechanical Engineering Department, Rajarshi Shahu College of Engineering, Tathawade, Pune, Maharashtra, India; Department of Automation & Robotics, Sharad Institute of Technology, College of Engineering, Yadrav, Ichalkaranji, Kolhapur, Maharashtra, India

V. S. Hajare
Department of Automation & Robotics, Sharad Institute of Technology, College of Engineering, Yadrav, Ichalkaranji, Kolhapur, Maharashtra, India

K. Shree Harini
Student of Electronics and Communication Engineering, PSG Institute of Technology and Applied Research, Coimbatore, India

S. B. Herwade
Department of Automation & Robotics, Sharad Institute of Technology, College of Engineering, Yadrav, Ichalkaranji, Kolhapur, Maharashtra, India

M. Jothibasu
Department of Electronics and Communication Engineering, PSG Institute of Technology and Applied Research, Coimbatore, India

Yogesh Kamble
Department of Automation & Robotics, Sharad Institute of Technology, College of Engineering, Yadrav, Ichalkaranji, Kolhapur, Maharashtra, India

Pankaj Kumar Keshri
School of Electronics and Electrical Engineering, Lovely Professional University, Punjab, India

S. N. Khan
Mechanical Engineering Department, Rajarshi Shahu College of Engineering, Tathawade, Pune, Maharashtra, India

Sanjay Khot
Sharad Institute of Technology, College of Engineering, Yadrav, Ichalkaranji, Kolhapur, Maharashtra, India

Usha Kiran
Department of Computer Science and Engineering, Gandhi Institute of Engineering and Technology University, Gunupur, Odisha, India

Abhishek Kumar
Associate Professor, School of Electronics and Electrical Engineering, Lovely Professional University, Phagwara, Punjab, India

Sujit Kumbhar
Department of Technology, Shivaji University, Kolhapur; Department of Automation & Robotics, Sharad Institute of Technology, College of Engineering, Yadrav, Ichalkaranji, Kolhapur, Maharashtra, India

Vaishali N. Mahavar
Aartav Engineering Private Limited, Gandhinagar, Gujrat, India

P. V. Miruthula
Student of Electronics and Communication Engineering, PSG Institute of Technology and Applied Research, Coimbatore, India

Chinmaya Kumar Palo
Department of Computer Science and Engineering, Gandhi Institute of Engineering and Technology University, Gunupur, Odisha, India

Govind Singh Patel
Department of Automation & Robotics, Sharad Institute of Technology, College of Engineering, Yadrav, Ichalkaranji, Kolhapur, Maharashtra, India

Raju Patel
School of Electronics Engineering, Vellore Institute of Technology, Chennai, India

Mohit Payal
Electronics and Communication Engineering, Graphic Era Hill University, Dehradun, Uttarakhand, India

G. V. Pujari
Department of Automation & Robotics, Sharad Institute of Technology, College of Engineering, Yadrav, Ichalkaranji, Kolhapur, Maharashtra, India

Ajay Roy
Professor, Department of Electrical Engineering, College of Engineering, University of Bisha, Bisha,Kingdom of Saudi Arabia

Shippu Sachdeva
School of Electronics and Electrical Engineering, Lovely Professional University, Phagwara, Punjab, India

Chita Ranjan Sahoo
ICMR-Regional Medical Research Centre, Department of Health Research, Ministry of Health and Family Welfare, Government of India, Bhubaneswar, India

Manoj Sindhwani
School of Electronics and Electrical Engineering, Lovely Professional University, Phagwara, Punjab, India

Balraj Singh
School of Computer Science Engineering, Lovely Professional University, Punjab, India

Dushyant Kumar Singh
Edunet Foundation, Gurgaon, India

Dhruv B. Trivedi
Civil Engineering Department, Rai University, Ahmedabad, Gujarat, India

Suman Turpati
Associate Professor, Department of ECE, Rajeev Gandhi Memorial College of Engineering and Technology, Nandyal, Andhra Pradesh, India

Anshu Vashisth
School of Computer Science Engineering, Lovely Professional University, Punjab, India

Foreword

The convergence of the Internet of Things (IoT) with sustainable practices has become a bright spot in the fast-evolving technology, pointing to a future in which environmental responsibility and innovation coexist harmoniously. The book, *IoT Potential for Green Energy Solution,* demonstrates the revolutionary potential of this partnership by illuminating the critical role that IoT plays in changing the face of green energy solutions. In light of the issues presented by climate change and the need to shift toward sustainable practices, we must investigate IoT applications within the green-energy domain. These chapters explore a variety of aspects of this intriguing intersection, providing a thorough examination of the enormous potential that IoT offers to completely transform how we think about energy distribution, consumption, and conservation.

"Sustainable Wind Energy: Optimizing Hybrid FRP Materials and Innovative Blade Design for Enhanced Efficiency" opens the discourse by dissecting advancements in wind energy, focusing on the integration of hybrid materials and cutting-edge blade designs. From the skies with "Introduction to Path Planning in an UAV Network" to the ground with "Commercial Fleet Sustainability: IoT Tracking Progress with Environmental Monitoring," the narrative unfolds, exploring the far-reaching effects of IoT in domains as diverse as agriculture, transportation, and energy storage. Chapters such as "Need-Based Irrigation Powered by IoT Towards Green Horticulture Cultivation" and "Empowering Smart Farming: Extensive Plant Surveillance and Accurate Pest Identification" underscore the pivotal role IoT plays in enhancing efficiency and sustainability in agriculture, while "Self-Powered IoT Devices: Harnessing Energy for a Connected Future" unveils the potential for self-sustaining IoT ecosystems.

Many angles are taken into consideration while examining how the Internet of Things is influencing the hybrid and renewable energy sector, including how it affects energy monitoring and management systems. The theme of "The Role of Internet of Things (IoT) in Hybrid and Renewable Energy" highlights the transformative power of interconnected technologies in crafting a more sustainable and environmentally friendly future.

In each chapter, the contributors bring forth a wealth of knowledge and insights, providing a holistic view of the intricate relationship between IoT and green energy. As we embark on this journey through the pages of *IoT Potential for Green Energy Solution*, may we be inspired by the innovative solutions presented within, and may these insights propel us toward a future where technology not only enriches our lives but also nurtures the very planet we call home.

—Dr. Somenath Chakraborty,
Assistant Professor
The West Virginia University Institute of Technology
Beckley, West Virginia
USA

Preface

The association of green energy solutions and the Internet of Things (IoT) presents an optimistic vision for a sustainable future in an era of accelerated technological advancement and climate change. The book *IoT Potential for Green Energy Solution* explores novel applications that hold the potential to completely transform the way energy is produced, distributed, and consumed. It accomplishes this by investigating the synergies between IoT and various aspects of renewable and hybrid energy systems. In an era where technological advancement and climate change are converging, it is critical to take advantage of the potential of the IoT to be the catalyst for the shift toward a more environmentally friendly and sustainable worldwide community. This book is an in-depth primer that offers insights into the immense potential that the Internet of Things has for influencing energy systems in the future.

—Editors

Introduction

The present book, *IoT Potential for Green Energy Solutions,* explores the integration of Internet of Things (IoT) technology and sustainable energy solutions. With a focus on environmental sustainability and green energy programs, this comprehensive book examines many elements of IoT integration in renewable energy systems. The book commences by elucidating the design methodology essential for sustainable energy components, while concurrently highlighting how IoT simplifies the comprehension of energy consumption patterns.

The chapters address a wide range of issues, beginning with "Sustainable Wind Energy: Optimizing Hybrid FRP Materials and Innovative Blade Design for Enhanced Efficiency." This section delves into improvements in wind-energy technology targeted at increasing efficiency through novel materials and blade designs. "IoT-Integrated Hybrid Renewable Energy System" looks into the integration of IoT technology in hybrid renewable energy systems with a focus on the smooth coordination and administration of several energy sources. The book also covers the use of IoT to address environmental concerns such as noise pollution, including chapters such as "The Use of the Internet of Things for Active Noise Control Technology." Other chapters look at IoT applications in a variety of industries, including transportation, commercial fleet sustainability, agriculture, and energy storage and distribution. "Need-Based Irrigation Powered by IoT Towards Green Horticulture Cultivation" emphasizes the role of IoT-powered irrigation systems in encouraging sustainable agriculture practices. Furthermore, the book looks at how the Internet of Things might improve energy monitoring, management, and conservation. "IoT-Enhanced Energy Storage and Distribution" investigates the potential of IoT technology to improve energy storage and distribution networks.

Overall, this book throws light on the revolutionary effects of IoT technology on the renewable energy sector, providing valuable perspectives for researchers, practitioners, and policymakers interested in promoting environmental sustainability and green energy initiatives.

CHAPTER 1

Sustainable Wind Energy: Optimizing Hybrid Fiber-Reinforced Polymer Materials and Innovative Blade Design for Enhanced Efficiency

ASHISH A. DESAI[1,2,*] and S. N. KHAN[1]

[1]*Mechanical Engineering Department, Rajarshi Shahu College of Engineering, Pimpri-Chinchwad, Maharashtra, India*

[2]*Department of Automation & Robotics, Sharad Institute of Technology, College of Engineering, Yadrav, Ichalkaranji, Kolhapur, Maharashtra, India*

*Corresponding author

ABSTRACT

The increasing need for sustainable wind energy necessitates new advancements in the composition and design of wind turbine blades. The objective of this study is to create stronger and stiffer blades by using hybrid fiber-reinforced polymers (FRP) composed of carbon fibers, bamboo fibers, and e-waste particles. ANOVA and Taguchi's L9 Orthogonal array design are used in the study to identify the primary determinants of mechanical features. The results demonstrate that the amount of fiber-reinforced e-waste greatly enhances functioning, while resin components have minimal effect. Therefore, it is beyond reasonable doubt that hybrid FRP composites have the potential to increase the sustainability and efficiency of wind energy systems. The study illustrates the ideal fiber kinds and blending methods for optimal stiffness and strength. Investigating durability and performance in the real world paves the way for developments in renewable energy.

IoT Potential for Green Energy Solutions. Abhishek Kumar, Ajay Roy, & Shamimul Qamar (Eds.)

DOI: 10.1201/9781779643599-1

Beyond wind energy, the findings affect other industries like aerospace and automotive because they make it possible to create stronger, lighter parts that are more efficient and emit fewer emissions. This study advances the wider use of sustainable composite solutions by improving material performance and lowering production costs. This study suggests a novel method for producing wind turbine blades that supports international initiatives to create a more sustainable and environmentally friendly world.

1.1 INTRODUCTION

Wind energy plays a significant role in the renewable energy landscape, but its future development is strongly reliant on material and design innovations. Wind power has expanded rapidly, ushering in the era of large-scale wind turbine blades. However, the intricate interaction of wind energy components in different geographical locations, as well as the benefits of these larger blades, remains a mystery. A thorough analysis of the weight, environmental implications, energy consumption, and cost variations induced by the usage of hybrid fiber-reinforced polymer (FRP) materials in wind turbine blades is required to address this issue [1–3].

This study studies the sustainability elements of wind energy, delving deeply into the properties of hybrid FRP materials and their use in cutting-edge wind turbine blade design. The capacity to alter these materials to satisfy the specific needs of different wind turbine designs and operational scenarios is a hot topic. Furthermore, the study investigates cutting-edge blade design approaches that employ hybrid FRP materials, elucidating how improved aerodynamic efficiency and decreased blade weight result in higher energy collection and structural durability. The study incorporates an emphasis on environmental sustainability, with a special emphasis on recyclability and minimizing environmental implications. Furthermore, the study emphasizes the crucial necessity of hybrid FRP materials and novel blade design approaches in boosting the long-term sustainability of wind energy. These improvements help to increase wind power efficiency, operating durability, and environmental commitment, seamlessly connecting wind energy with the global search for renewable and sustainable energy sources. In essence, the essay is an excellent resource for researchers, scientists, and technicians, as well as decision-makers, who are enthusiastic about advancing the future of wind energy while understanding its complicated environmental implications [1].

1.2 LITERATURE REVIEW

1.2.1 AN OVERVIEW ON THE IMPORTANCE OF WIND ENERGY AS A SUSTAINABLE ENERGY SOURCE

Wind energy is a sustainable and renewable energy source that provides long-term reliability because it is based on the constant motion of the Earth's atmosphere caused by the sun's energy. Because it is based on the constant motion of the Earth's atmosphere driven by the sun's energy, wind energy is a sustainable and renewable energy source with long-term reliability. Unlike finite fossil fuels, the wind will continue to blow, providing essentially infinite wind energy. These publications support the value of wind energy as a sustainable energy source. It has prioritized large-scale offshore wind farms as critical to reducing greenhouse gas emissions and achieving energy independence, as well as a broad assessment of wind energy while emphasizing the broad nature of the engineering difficulties they imply. Recent advances in wind energy technology are investigated, and the study focuses on the major obstacles in designing and building wind energy systems [2]. It also emphasizes the serious challenges faced by climate change, as well as the growing recognition of this issue among leaders in the public and private sectors. The path forward includes a comprehensive reform of the global energy system, including the replacement of fossil fuels with renewable energies such as wind power, as well as significantly increased energy efficiency. This transition is crucial for achieving sustainability and mitigating the effects of climate change [3]. It emphasizes the significance of doing comprehensive wind energy assessments in order to maximize energy gathering. Overall, these papers demonstrate the value of wind energy in combating climate change, promoting sustainability, and reducing reliance on nonrenewable resources [4–6].

1.2.2 A DESCRIPTION OF THE PROBLEM OR RESEARCH GAP IN WIND TURBINE EFFICIENCY THROUGH MATERIAL SELECTION

While traditional nonrenewable resource use is effective, it causes challenges with wind turbine efficiency due to material selection. Traditional materials, such as those used in the manufacture of wind turbines, are frequently produced from scarce and environmentally dangerous sources. This not only calls into doubt the long-term viability of the wind energy sector but also underscores the need for more environmentally friendly alternatives. There is a research gap in developing materials that can perform similarly to or better than existing solutions while being sustainable and cost-effective. Innovative materials that balance performance, sustainability, and cost-effectiveness are an important area of study for overcoming this challenge in wind turbine efficiency through material selection [14].

The selection of materials is critical in determining the overall performance of wind turbines. Nonrenewable materials have traditionally been used in wind blade production; however, concerns have developed due to their limited supply and negative environmental impact. As a result of these issues, there is a rising need to investigate alternatives that are both sustainable and cost-effective. This search for new materials is critical for increasing wind turbine efficiency while reducing the industry's negative environmental impact. The development of sustainable and economically viable alternatives reflects a broader commitment to promoting greener wind energy technologies [7]

1.2.3 HYBRID FIBER-REINFORCED COMPOSITES MATERIAL POTENTIAL WITH INNOVATIVE WIND TURBINE BLADES

Wind turbine blades are exposed to adverse climatic conditions such as strong winds, temperature variations, and UV exposure. Traditional materials like glass fiber-reinforced polymers (GFRP) have served as the foundation of blade construction. However, hybrid FRP materials have been developed due to the limits of GFRP in terms of weight, strength, and durability. These materials blend various fibers, such as carbon, glass, and natural fibers, with different resin matrices, resulting in improved mechanical qualities and increased sustainability.

For wind turbine blades, hybrid FRP materials provide various advantages. They can provide a good blend of stiffness and toughness, lowering the danger of damage from high loads and fatigue. Hybridization also allows for the optimization of material properties, resulting in lightweight and robust blades [20]. Wind turbine blade design innovation focuses on making blades that are more efficient, aerodynamic, and robust. Advanced aerodynamic profiles, such as airfoil forms influenced by aviation technology, twisting blades, and tip serrations, are frequently used in these designs. The objective is to maximize wind energy capture while minimizing structural loads and noise. In addition, active load management technologies are being incorporated into the blades to enable real-time adjustments for enhancing energy generation and diminishing the effects of wind gusts and turbulence. Innovative blade designs are essential for increasing wind turbine efficiency and lowering maintenance costs. This study focuses on the advancement of wind energy systems, with a particular emphasis on wind turbines. Wind energy is widely acknowledged as an important renewable energy source, and optimizing its output and utilization is critical for a sustainable energy future [8].

Various authors offer insights into the usage of glass/carbon fiber, natural fiber, and e-waste as potential alternatives to traditional composite materials in wind turbine applications. It will look into the mechanical qualities, environmental impact, and cost-effectiveness and scalability of these materials. Researchers can provide valuable information to the wind energy industry by examining the performance of alternative materials through different testing methods. This analysis can help inform decisions on material selection and design optimization [10].

Overall, by optimizing wind energy systems through unique blade design and the use of hybrid FRP materials, it may maximize energy output while minimizing the environmental impact of energy production.

1.3 METHODOLOGY

The production of hybrid composite samples is a step-by-step process that begins with material selection and ends with extensive testing. To begin with woven carbon fibers with a thickness defined in grammas per square meter are selected to achieve the appropriate strength and stiffness. An epoxy resin is carefully chosen based on its compatibility with carbon fibers and intended use, taking into account parameters such as temperature resistance and cure time. The binder composition is optimized to provide compatibility with e-waste particles, enhancing composite homogeneity. The e-waste particle grain size is designed to provide uniform distribution and to meet sustainability objectives. A compression-assisted hand layup technique is used in the production process, in which layers of carbon fibers, epoxy glue, and e-waste particles are precisely assembled in a specified order. It is critical to achieve good bonding and consolidation of composite layers in order to ensure the structural integrity and performance of composite materials. The manual compression method is a process for pressing or squeezing the layers of composite materials together, allowing for good bonding and consolidation. Following manufacture, the samples are subjected to rigorous testing to determine their mechanical qualities and overall performance. Tensile and flexural test reveal information about the hybrid composite's strength, stiffness, and durability. Visual inspections are performed to detect any flaws or discrepancies. The results of these tests lead an iterative optimization process that allows for changes in parameters such as fiber thickness or e-waste particle size to improve the performance of the composite [11–13].

Finally, the progressive manufacture of hybrid composite samples, combined with extensive testing, ensures the development of materials with specific properties. This method not only improves the composite's efficiency and sustainability, but it also contributes to the overall improvement of composite manufacturing technologies for applications in renewable energy and other industries. The methodology is being considered as illustrated in Figure 1.1.

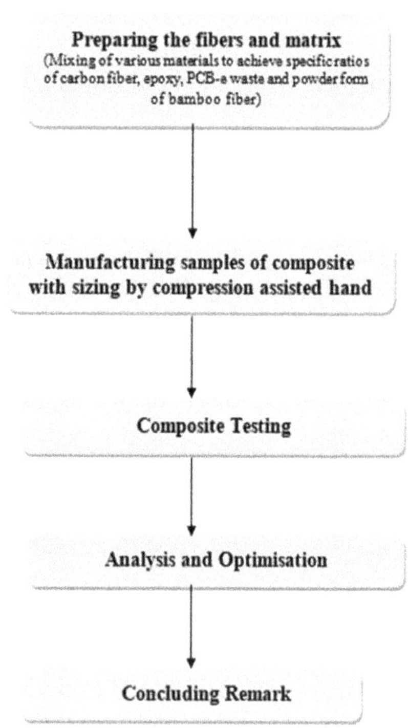

FIGURE 1.1 Working methodology.

1.3.1 WIND BLADE MATERIAL SELECTION AND PREPARATION

1.3.1.1 CARBON FIBER AS A REINFORCEMENT MATERIAL IN WIND ENERGY APPLICATIONS

Carbon fiber is well-known for its high strength-to-weight ratio, which makes it an excellent choice for wind turbine blades. Its high rigidity allows for efficient energy transmission, which leads to increased turbine performance. Furthermore, the corrosion resistance of carbon fiber maintains the blades' lifespan. However, the high cost of carbon fiber and the complex production process make widespread adoption for wind energy applications difficult. Furthermore, because carbon fiber composites are nonbiodegradable, their disposal raises environmental concerns. Exploring alternate materials and creating more efficient manufacturing procedures could aid in mitigating the high cost and complexity of carbon fiber. Furthermore, creating means to recycle carbon fiber composites or finding sustainable ways to dispose of them could address environmental problems.

1.3.1.2 BAMBOO AS AN ALTERNATIVE COMPOSITE MATERIAL

Bamboo has emerged as a promising alternative composite material for overcoming carbon fiber's shortcomings. It is a renewable and abundant resource that may be obtained in a sustainable manner, alleviating environmental concerns. Bamboo composites have demonstrated remarkable mechanical qualities, such as a high strength-to-weight ratio and fatigue resistance, making them a feasible option for reducing turbine weight and enhancing wind energy efficiency. Furthermore, the cheaper production costs and simpler manufacturing procedures of bamboo compared to carbon fiber may help reduce the high cost and complexity involved with employing composites in wind turbine building.

1.3.1.3 PCB E-WASTE MATERIAL

The inclusion of printed circuit board (PCB) e-waste materials into wind blade manufacture in the form of composite powder is a promising prospect for a more environmentally responsible and cost-effective solution. This novel strategy involves replacing traditional PCB materials with bamboo composites, addressing both economic and environmental problems in the wind turbine sector. The traditional disposal of PCB materials in wind turbine manufacture has been linked to the generation of electronic waste and environmental damage. The industry can address these concerns by switching to bamboo composites, encouraging sustainability and lowering its environmental effect. This transition is an important step toward a more environmentally friendly and economically viable future for wind energy technologies.

1.3.2 COMPOSITE PREPARATION

Following the assessment of the literature, specific materials, and weight percentages were identified for further analysis and incorporation into the design parameters. Impact, percent weight of carbon fiber, percent electronic waste (especially PCB material), and powder form of bamboo are among the materials chosen. These are essential design variables that will be thoroughly explored in order to appreciate their individual and combined effects on the composite material's properties and performance. The precise formulation and combining of various components to create particular ratios of carbon fiber, epoxy, bamboo fiber in powder form, and PCB e-waste is required for the preparation of composite samples.

1.4 EXPERIMENTAL RANGE AND LEVELS OF INDEPENDENT VARIABLES

These properly identified design factors provide a comprehensive approach to material optimization, with each element playing a distinct function in increasing mechanical performance, resolving environmental issues, and examining sustainable alternatives. The study's goal is to decode the complex interactions and influences of numerous design factors on composite material properties in order to inform a well-balanced and purpose-driven formulation, which is formulated in Table 1.1.

As illustrated, the compression-assisted hand lay-up technique improves the precision of fiber/matrix stacking while minimizing voids inside the composite material. Sheet boxes measuring ASTM 3039 D, 250mm long × 25 mm broad × 2 mm thickness, were used in this manner, as shown in Figure 1.2.

TABLE 1.1 Experimental Range and Levels of Independent Variables

Level	Wt.% Carbon Fiber	E-waste (%)	Bamboo Fiber (%)
1	10	5	2
2	20	10	5
3	30	15	10

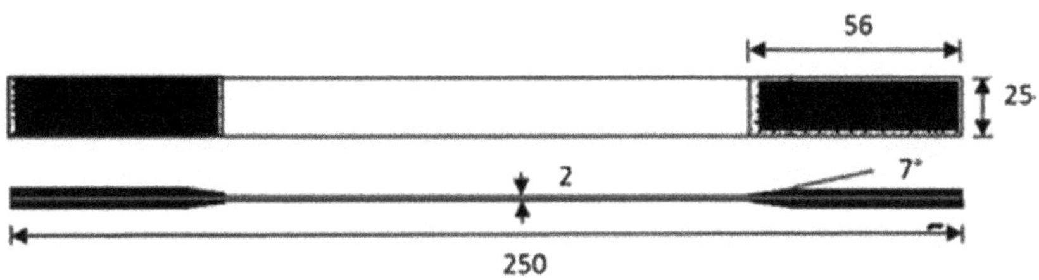

FIGURE 1.2 ASTM standard specimens for tensile and flexural testing.

This approach involves manually putting up the reinforcing fibers and matrix, as well as using compression to provide a more uniform distribution of materials and reduce the possibility of void development inside the composite structure. The sheet box dimensions demonstrate the controlled and uniform conditions employed during the molding process, which helps to ensure the uniformity and dependability of the final tensile and flexural specimens. Figure 1.3 illustrates the tensile and flexural specimens as well as the experimental setup.

1.5 RESULTS AND DISCUSSIONS

The Taguchi L9 orthogonal array was employed in the experimental design for this investigation, giving a systematic and efficient method for measuring numerous variables influencing wind turbine blade performance (see Table 1.2).

The study utilized the Taguchi L9 orthogonal array to identify the most effective parameters for enhancing tensile and flexural strength. This systematic and controlled experimental approach enables researchers to draw informed conclusions about the key factors influencing mechanical performance. The results of the Taguchi L9 array analysis can be used to guide further steps in the optimization process, resulting in enhanced product or material attributes. Overall, the Taguchi L9 orthogonal array is a powerful tool for effectively unraveling the intricate correlations between variables and mechanical qualities, supporting the quest of improved product performance in engineering and industrial applications.

FIGURE 1.3 Setup for the experiment: (a) tensile test (b) flexural test.

TABLE 1.2 Experimental Setup Results for Tensile Strength and Flexural Strength

Exp. No.	Wt.% of Carbon Fiber	% of Bamboo Fiber	% of E-Waste	Tensile Strength (MPa)	Flexural Strength (MPa)
1	10	5	2	157	91
2	10	10	5	178	125
3	10	15	10	185	163
4	20	5	5	176	134
5	20	10	10	196	140
6	20	15	2	196.3	155
7	30	5	10	183	156
8	30	10	2	310	203
9	30	15	5	316.4	208

1.5.1 EXPERIMENTAL DESIGN

Statistical data analysis approaches, including the Taguchi method, were used to investigate the effect of composite weight percentage on mechanical parameters. The signal-to-noise (S/N) ratio and analysis of variance (ANOVA) were both used in the analysis. These techniques are commonly used in experimental design and statistical analysis to study the effect of variables on response variables. This efforts' ultimate goal is to maximize S/N ratio, as defined by the Taguchi technique.

$$S/N = -10 * \log (\Sigma(1/Y^2)/n)$$

TABLE 1.3 Experimental Results of Tensile Strength with a Corresponding S/N Ratio and Mean 1

Exp. No.	wt.% of Carbon Fiber	% of Bamboo Fiber	% of E-waste	Tensile Strength (MPa)	SNRA1	MEAN 1
1	10	5	2	157	43.91799	157
2	10	10	5	178	45.0084	178
3	10	15	10	185	45.34343	185
4	20	5	5	176	44.91025	176
5	20	10	10	196	45.84512	196
6	20	15	2	196.3	45.85841	196.3
7	30	5	10	183	45.24902	183
8	30	10	2	310	49.82723	310
9	30	15	5	316.4	50.00473	316.4

Figures 1.4 and 1.5 show the main effect plot for SN ratios *versus* tensile strength and the main effect plot for means *versus* tensile strength, which are used to discover patterns and relationships that would not be visible from numerical data alone (see Tables 1.3 and 1.4).

The main effect plots for SN ratios–flexural strength and means–flexural strength are shown in Figures 1.6 and 1.7. To identify the ideal combination for wind turbine blade design, the Design of Experiments (DOE) can be used to systematically adjust factors such as carbon fiber, epoxy, e-waste, and bamboo fiber powder. DOE allows for the investigation of their individual and combined effects on blade performance by running tests with varying values of these parameters.

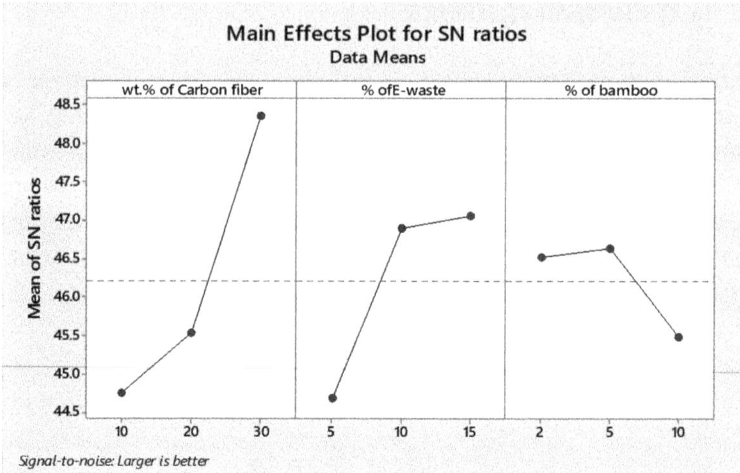

FIGURE 1.4 Main effect plot for SN ratios–tensile strength.

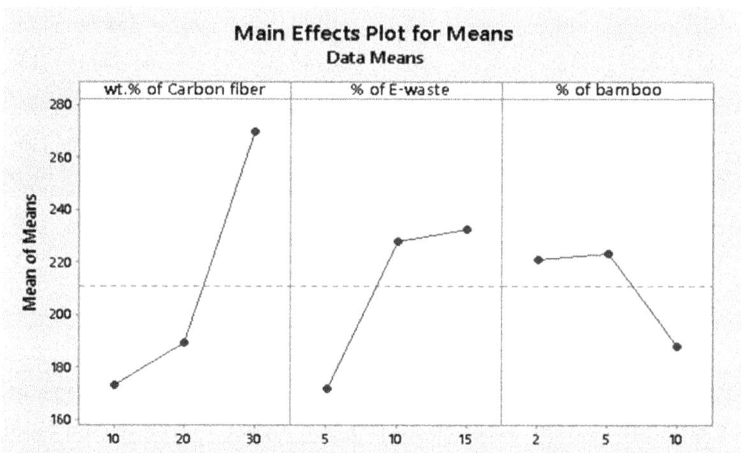

FIGURE 1.5 Main effect plot for means–tensile strength.

1.5.2 RESPONSE TABLE FOR SIGNAL-TO-NOISE RATIOS FOR TENSILE STRENGTH AND FLEXURAL STRENGTH

A Response Table for S/N Ratios is a useful tool in the Taguchi approach and other experimental procedures for tensile and flexural strength. Tables 1.5 and 1.6 show the ANOVA test results for tensile and flexural strength.

TABLE 1.4 Experimental Results of Flexural Strength with a Corresponding S/N Ratio and Mean 2

Exp. No.	wt.% of Carbon Fiber	% of E-waste	% of Bamboo	Flexural Strength (MPa)	SNRA2	MEAN2
1	10	5	2	91	39.18083	91
2	10	10	5	125	41.9382	125
3	10	15	10	163	44.24375	163
4	20	5	5	134	42.5421	134
5	20	10	10	140	42.92256	140
6	20	15	2	155	43.80663	155
7	30	5	10	156	43.86249	156
8	30	10	2	203	46.14992	203
9	30	15	5	208	46.36127	208

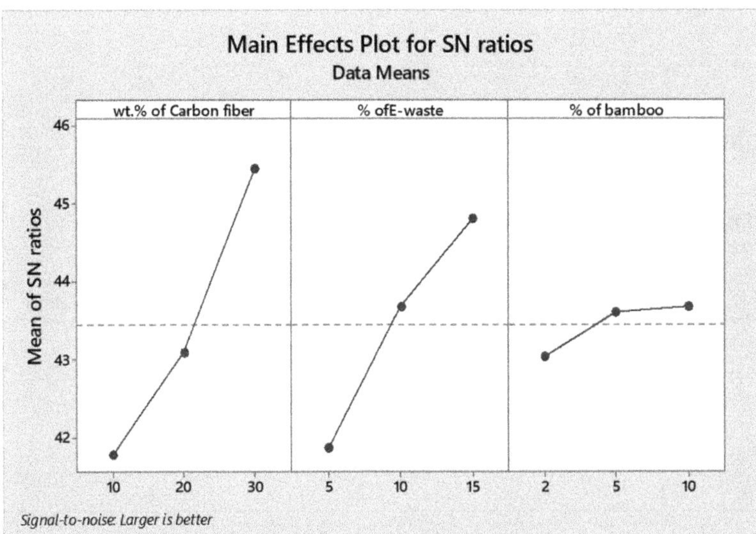

FIGURE 1.6 Main effect plot for SN ratios–flexural strength.

1.5.3 *ANALYSIS OF VARIANCE FOR S/N RATIOS*

Based on the analysis of variance (ANOVA) in Table 1.7, the weight percentage of carbon fiber and the percentage of e-waste utilized emerged as the primary design factors influencing the tensile strength of hybrid composite laminates. Specifically, carbon fiber weight percentage was found to account for 57.04% of the variance, while the percentage of e-waste contributed 27.83%.

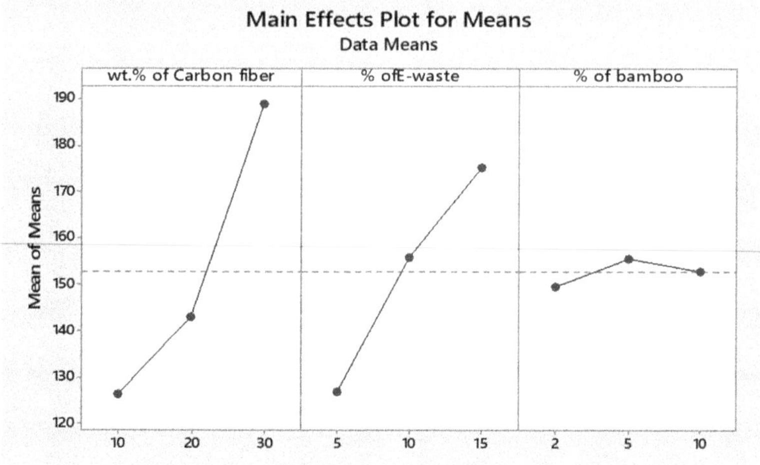

FIGURE 1.7 Main effect plot for means–flexural strength.

TABLE 1.5 Response Table for Signal-to-Noise Ratios for Tensile Strength

Larger Is Better			
Level	**Wt.% of Carbon Fiber**	**% of E-waste**	**% of Bamboo**
1	44.76	44.69	46.53
2	45.54	46.89	46.64
3	48.36	47.07	45.48
Delta	3.60	2.38	1.16
Rank	1	2	3

TABLE 1.6 Response Table for Signal-to-Noise Ratios for Flexural Strength

Larger Is Better			
Level	**Wt.% of Carbon Fiber**	**% of E-waste**	**% of Bamboo**
1	41.79	41.86	43.05
2	43.09	43.67	43.61
3	45.46	44.80	43.68
Delta	3.67	2.94	0.67
Rank	1	2	3

TABLE 1.7 Analysis of Variance for S/N Ratios for Tensile Strength

Source	DOF	Seq. SS	Adj SS	Adj MS	F	P	% C	
wt.% of Carbon fiber	2	21.563	21.563	10.782	6.66	0.131	57.04	Significant
%of E-waste	2	10.523	10.523	5.262	3.25	0.235	27.83	Significant
%of bamboo	2	2.475	2.475	1.238	0.76	0.567	6.54	Significant
Residual error	2	3.240	3.240	1.620				
Total	8	37.801						
R-Sq = 89.90% R-Sq(adj) = 59.59%								

In conclusion, the ANOVA analysis demonstrated that the wt.% of carbon fiber and the percentage of e-waste had a significant impact on the flexural strength of the hybrid composite laminates utilized in wind turbine blade design (see also Table 1.8.)

TABLE 1.8 Analysis of Variance for S/N ratios for Flexural Strength

Source	DOF	Seq. SS	Adj SS	Adj MS	F	P	% C	
wt.% of Carbon fiber	2	20.7734	20.7734	10.3867	5.77	0.148	54.23	Significant
% of E-waste	2	13.2114	13.2114	6.6057	3.67	3.67	34.48	Significant
% of Bamboo	2	0.7241	0.7241	0.7241	0.20		1.889	
Residual error	2	3.5972	3.5972	3.5972				
Total	8	38.3061						
R-Sq = 90.6% R-Sq(adj) = 62.4%								

1.5.4 INTERACTION GRAPH

Interaction graphs are graphic representations of relationships in the wind power industry. Graphs in Figures 1.8 and 1.9 frequently depict a response variable against one component, with many lines or curves representing varying amounts of the other factor. These interaction graphs are used to examine material interactions and discover how different materials affect wind turbine performance.

Based on these figures, ideal material combinations may be discovered that produce the highest strength, lightest weight, and longest durability. This insight is crucial in the design of windmills that maximize energy generation while minimizing environmental impact, in line with renewable energy sustainability goals. Figures 1.10 and 1.11 show an interaction plot of data means for the weight percent of carbon fiber and the weight percent of bamboo fiber for flexural strength.

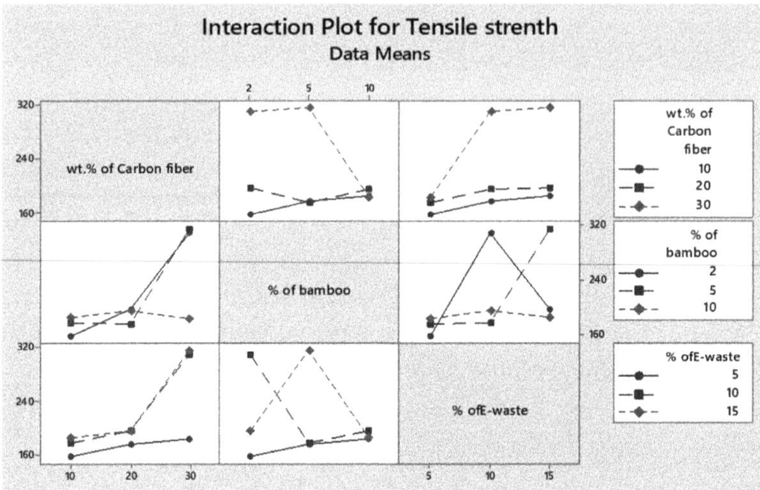

FIGURE 1.8 Interaction plot of data means for wt.% of carbon fiber, % of bamboo, and % of e-waste for tensile strength.

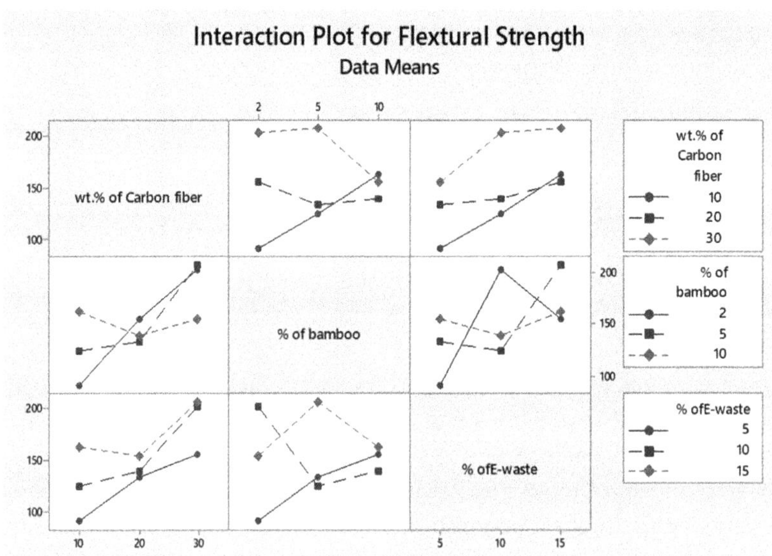

FIGURE 1.9 Interaction plot of data means for wt.% of carbon fiber, % of bamboo, and % of e-waste for flexural strength.

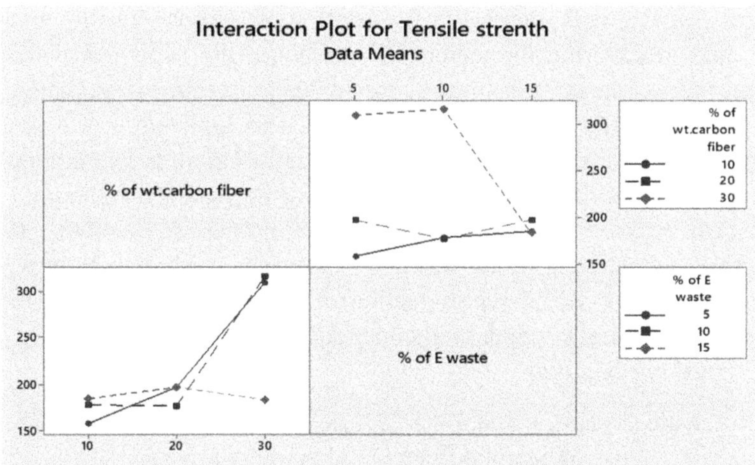

FIGURE 1.10 Interaction plot of data means for wt.% of carbon fiber and % of e-waste for tensile strength.

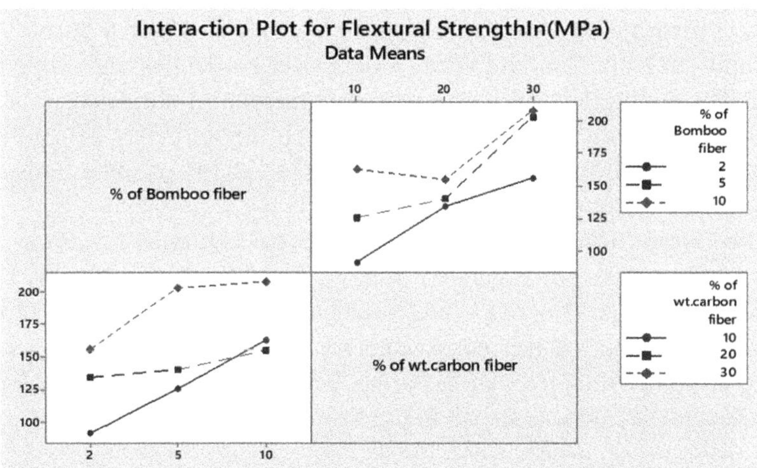

FIGURE 1.11 Interaction plot of data means for wt.% of carbon fiber and % of e-waste for tensile strength normal probability for flexural strength (MPa).

1.6 LINEAR REGRESSION ANALYSIS

Linear regression analysis is a versatile and widely used statistical approach for modeling and analyzing the relationships between variables.

It provides a formal framework for creating predictions, testing hypotheses, and discovering the factors that influence the dependent variable. Linear regression analysis allows researchers to examine how changes in one variable influence changes in another. The application of a normal probability plot for tensile and flexural strength is vital in determining the normality of the data distribution, providing important insights into the statistical behavior of these critical material properties. Similarly, mean effect plots generated by statistical software such as Minitab 17 provided a comprehensive visual representation of the significant differences in tensile and flexural strength seen under different test conditions.

Regression Equation:

Tensile strength = 108.4 + 4.82% of wt. carbon fiber + 6.89% of bamboo fiber + 3.31% of e-waste

Flexural strength in (MPa) = 53.8 + 3.133% of wt. carbon fiber + 5.82% of bamboo fiber + 0.33% of e-waste

A normal probability plot is a useful tool for determining the normality of data distributions, particularly in the context of material tensile and flexural strength. This graphical tool assists academics and engineers in understanding how well observed data matches a normal (Gaussian) distribution. Many statistical studies rely on the normalcy assumption, and deviations from normality may affect the validity of some statistical tests.

The term "mean effect plots generated by Minitab 17" means that mean effect plots are generated using statistical software, in this case Minitab 17. Mean effect plots are graphical representations of the impact of several factors on the mean response. Changes in attributes such as material composition or manufacturing factors might alter average tensile and flexural strength, as shown in Figures 1.12 and 1.13.

1.7 CONCLUSION

The findings of the study demonstrate that hybrid composites exhibit significantly greater tensile and flexural strength compared to traditional carbon fiber alternatives. By combining carbon fiber and bamboo components, the overall strength of the composites is notably enhanced. In contrast, while at present, the e-waste component has a minor impact on material strength. The strategic application of the Taguchi

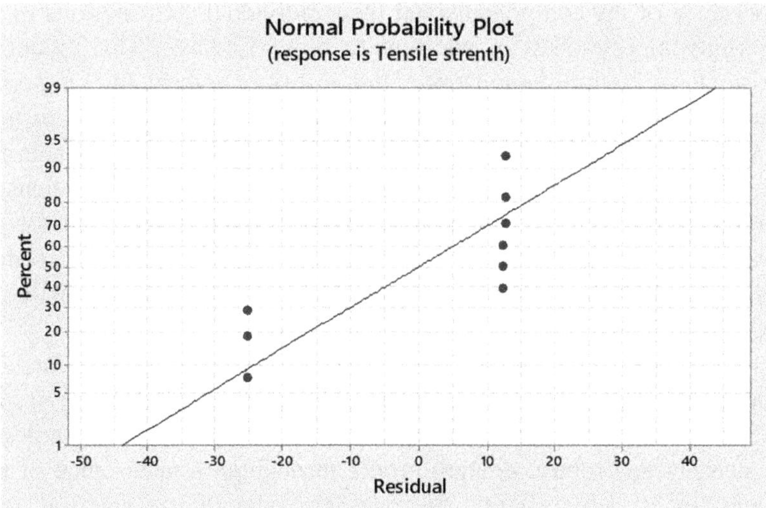

FIGURE 1.12 Normal probability for tensile strength (MPa).

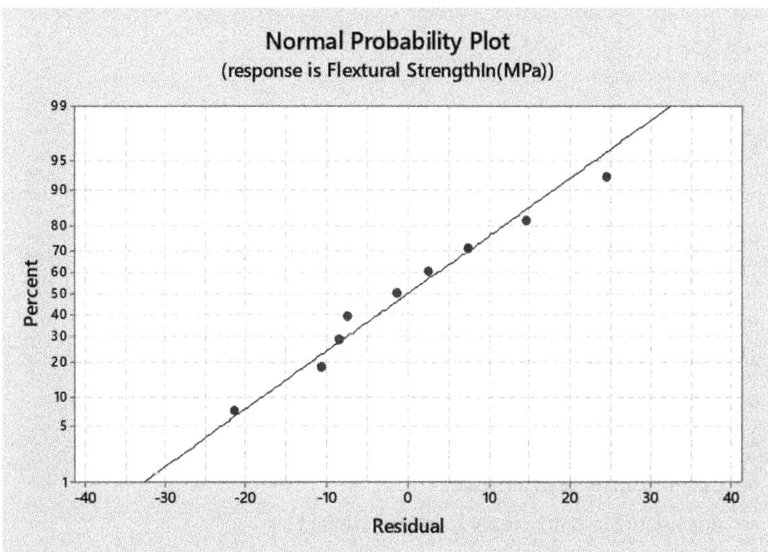

FIGURE 1.13 Normal probability for flexural strength (MPa).

L9 orthogonal array design allows for the identification of appropriate fiber weight ratios, offering critical insights for improving the hybrid composites manufacturing process. ANOVA analysis validates the importance of the components and their individual contributions to the experimental responses in addition to these findings. This extensive statistical validation demonstrates the ability of hybrid FRP materials to improve the sustainability and efficacy of wind energy, establishing them as a promising solution in the field of renewable energy sources. The results of our research not only contribute to our comprehension of material performance but also offer actionable suggestions for the advancement and enhancement of wind turbine technologies. According to the study's findings, hybrid composites offer much higher tensile and flexural strength than regular carbon fiber competitors.

Finally, the data demonstrate the transformative influence of merging hybrid FRP materials with innovative blade design on wind energy system efficiency. The statistical examination of material properties indicates a consistent and robust strength profile throughout a wide range of test circumstances, demonstrating the materials' reliability in practical, real-world applications. This knowledge is essential for assuring the overall performance and dependability of wind turbine blades, hence improving the sustainability and effectiveness of wind energy as a renewable energy source.

KEYWORDS

- **sustainable wind energy**
- **hybrid FRP materials**
- **innovative blade design**
- **wind turbine**
- **efficiency**
- **renewable energy optimization**
- **eco-friendly wind power**
- **sustainable materials in wind turbines**
- **green energy solutions**
- **wind turbine innovation**
- **eco-efficient blade design.**

REFERENCES

1. Bai, C.-J. & Wang, W.-C. Review of computational and experimental approaches to analysis of aerodynamic performance in horizontal-axis wind turbines (HAWTs). *Renewable and Sustainable Energy Reviews, Elsevier* **2016**, *63*(C), 506–519.
2. Georgios, N., Bhattacharya, S., Vimalan, N., Demirci, H. E., Nikitas, N., & Kumar, P. Wind power: A sustainable way to limit climate change. *Managing Global Warming* **2019**, 333–364.
3. Kaygusuz, K. Wind power for a clean and sustainable energy future. *Energy Sources, Part B: Economics, Planning, and Policy* **2009**, *4*(1), 122–133.
4. Herbert, G. M. J., Iniyan, S., Sreevalsan, E., & Rajapandian, S. A review of wind energy technologies. *Renewable and Sustainable Energy Reviews* **2007**, *11,* 1117–1145
5. Alcayde, A., Hernandez-Escobedo, Q., Muñoz-Rodríguez, D., & Perea-Moreno, A.-J. Trends on optimizing wind turbine efficiency. *Worldwide Research Energies* **2022**, *15*, 6559.
6. Sawant, M., Thakare, S., Prabhakara Rao, A., Feijóo-Lorenzo, A. E., & Bokde. N. D. A review on state-of-the-art reviews in wind-turbine- and wind-farm-related topics. *Energies, MDPI,* **2021**, *14*(8), 1–30l.
7. Hayman, B., Wedel-Heinen, J., & Bmndsted, P. Material challenges in present and future wind energy. *MRS Bull* **2008**, *33*(4), 343–353.
8. Belfkira, Z., Mounir, H., & El Marjani, A. Structural optimization of a horizontal axis wind turbine blade made from new hybrid composites with kenaf fibers. *Composite Structures* **2020**, *260*, 113252.
9. Ning, S. A., Damiani, R., & Moriarty, P. J. Objectives and constraints for wind turbine Optimization. *Journal of Solar Energy Engineering* **2014**, *136*, 041010. doi: 10.1115/1.4027693.
10. Krzyżak, A., Relich, S., Kosicka, E., Szczepaniak, R., & Mucha, M. Selected construction properties of hybrid epoxy composites reinforced with carbon fiber and alumina. *Advances in Science and Technology Research Journal* **2022**, *16*, 240–248.
11. Samuel, B. O., Sumaila, M., & Dan-Asabe, B. Multi-objective optimization and modeling of a natural fiber hybrid reinforced composite (PxGyEz) for wind turbine blade development using grey relational analysis and regression analysis. *Mechanics of Advanced Materials and Structures* **2022**, *31*(1), 1–19.
12. Samuel, B. O., Sumaila, M., & Dan-Asabe, B. Modeling and optimization of the manufacturing parameters of a hybrid fiber reinforced polymer composite PxGyEz. *International Journal of Advanced Manufacturing Technology* **2022**, 1–12.
13. Samuel, B. O. Modeling and optimization of the manufacturing parameters of a hybrid fiber reinforced polymer composite PxGyEz. *International Journal of Advanced Manufacturing Technology* **2022**, *118*, 1441–1452.
14. Edeerozey, A. M., Akil, H. M., & Azhar, A. B. Chemical modification of kenaf fibres. *Materials Letters* **2007**, *61*, 2023–2025.

CHAPTER 2

IoT-Integrated Hybrid Renewable Energy System

BISWARANJAN BHOLA[1,*], CHINMAYA KUMAR PALO[1], USHA KIRAN[1], and CHITA RANJAN SAHOO[2]

[1]*Department of Computer Science and Engineering, Gandhi Institute of Engineering and Technology University, Gunupur, Odisha, India*

[2]*ICMR-Regional Medical Research Centre, Department of Health Research, Ministry of Health and Family Welfare, Government of India, Bhubaneswar, Odisha, India*

*Corresponding author

ABSTRACT

The shift toward more sustainable energy solutions has been driven by the increasing global is concerned about climate change, energy security, and reliance on fossil fuels. In order to overcome these challenges, this chapter investigates the incorporation of hybrid renewable energy systems, with an emphasis on options for electricity generation and fuel delivery. With a focus on wind energy the possibilities of several renewable sources for the generation of green hydrogen, including wind, solar, biomass, geothermal, and small hydropower. Lowering the impact on the environment is dependent on energy saving via the use of efficient construction materials and technology. Additionally, smart grids, IoT, and big data help Home Energy Management Systems, provide innovative methods to optimize household energy use. Real-time monitoring, predictive maintenance, and grid optimization are made possible by IoT, which enhances dependability and energy efficiency. A foundation for economical, environmentally friendly energy systems is provided by the suggested IoT architecture for

IoT Potential for Green Energy Solutions. Abhishek Kumar, Ajay Roy, & Shamimul Qamar (Eds.)

DOI: 10.1201/9781779643599-2

smart grid technologies and analytical models. Reliance on nonrenewable resources is decreased and environmental harm is minimized when renewable energy sources are integrated into a coherent system. With 78.96% forecast accuracy, smart grids provide an intelligent and effective way to manage energy in the future.

2.1 INTRODUCTION

The recurrence of global energy security concerns highlights oil dependency, geopolitical instability, and environmental issues. It provides a measured model of supply, demand growth, and price fluctuations to measure energy security costs come into play. The context is South Korea's energy policy shift toward sustainability [1]. The world's main energy challenge is replacing limited crude oil and reducing greenhouse gas emissions. Conventional policies see nuclear, fossil fuels, and renewable energy as competitors. However, the evolving energy landscape dictates that these resources must be strictly integrated to meet the needs of society. The work explores alternatives, focusing on liquid fuel and electricity generation [2].

Energy conservation, an essential part of the national strategy, can reduce environmental impact and fuel consumption. Insulation materials such as expanded polystyrene and rock wool provide energy storage in buildings with a couple of years of payback. The focus is on reducing heating costs and environmental impact [3]. The transition from fossil fuels to renewable energy is a global need. The primary solution for sustainable energy and storage continues to be hydrogen, derived from a range of materials. The method established by hydroelectrolysis produces high-purity hydrogen. Integrating electrolysis into renewable processes shows potential, although economic viability remains challenging [4]. This study examines the potential of renewable energy sources such as wind, solar, biomass, public waste, geothermal energy, and small hydropower for green hydrogen production. It highlights wind energy as a promising strategy for green hydrogen production [5].

Home energy management systems (HEMS) propose a comprehensive model; the research focuses on HEMS systems, communication systems, change management, software, and user engagement [6]. The proposed model integrates elements from existing models and incorporates external factors to actively involve end users to improve household energy efficiency and create demand for new systems. The basic focus is on developing the smart home, emphasizing energy efficiency and remote control [7]. It discusses microcontroller design and its role in controlling home subsystems. However, the presented text does not directly address the issue of converting power systems from home to commercial modules.

The development of energy systems, identifying decentralized, centralized, distributed, and intelligent energy systems, introduces the fundamental concepts of the Internet of Energy, such as prosumers, virtual power plants, microgrids, smart grids, and smart energy [8]. Performance standards and assessment functions for Energy Internet are described. A home smart energy management system using IoT and Big Data meshes a wireless network of devices, collecting energy data and transmitting it to a centralized server [9]. They use BI and Big Data analytics to improve energy efficiency, focusing on heating, ventilation, and air conditioning systems as case studies. IoT plays a significant role in revolutionizing smart energy systems across various industries including manufacturing, energy services, data networks, and electricity generation [10]. It addresses connectivity, complexity, and data volume challenges while focusing on security and privacy concerns in IoT applications. The study highlights potential volatility and associated risks in the energy industry.

IoT plays a key role in energy by enabling real-time monitoring and control of generation, distribution, and consumption [11]. Smart sensors and devices facilitate resource efficiency, grid optimization, predictive maintenance, and demand response, increasing energy efficiency and reliability. It plays a crucial role in transforming energy systems by improving real-time analysis, distribution, and facilitating energy efficiency. Connecting devices and enabling data-driven decisions contribute to energy efficiency and continuous improvement in manufacturing, distribution, and analysis systems [12]. Smart grids (SGs) provide two-way communication, real-time monitoring of devices, and energy efficiency. IoT enables SGs to balance generation and consumption better, improving reliability and energy conservation [13]. IoT devices such as sensors, actuators, and smart meters enhance real-time data collection and monitoring, enabling energy efficiency and grid management and improving energy systems reliability and sustainability [14, 15].

It focuses on building an integrated framework for Big Data analytics in IoT and Smart City applications. It aims to leverage data sources to create valuable analytics services for urban decision-makers with multiple data sources, and an integrated system is needed to facilitate service development [16, 17]. IoT-enabled data for machine learning in cloud, edge, and fog systems with a focus on current challenges. It proposes a flexible learning framework for IoT applications [18, 19]. The importance of energy systems, their management using IoT, and their worth can be analyzed using Big Data, cloud, and machine learning techniques.

2.1.1 DIFFERENT ENERGY SOURCES

The solution to increasing energy resource demand is the pressing issue of sustainable energy, driven by declining fossil fuels and rising global energy demand. It should be abundant, accessible, environmentally friendly, and climbable. One important energy source is (1) solar energy, a promising renewable resource that meets the world's growing energy demand. Its affordability and increasing growth make it a key solution, especially for developing countries, which may be an efficient solution for the global energy crisis [20–22]. (2) Another significant energy source is wind energy conversion system control strategies, especially pitch angle control, are important in energy production efficiency. The detailed analysis is a valuable model for further research in wind energy control systems [23].

The study examines wind power generation costs in Europe, with onshore wind generation costs of 4.5–8.7 €cents/kWh, offshore costs of €6–11.1/kWh, and total published hours and capital costs significantly impacting despite a 20% cost increase over the past three years due to rising input prices, wind power remains competitive, with the potential to reduce costs through research, development, maintenance, and planning in the future. Hydropower generated from gravity-driven [24] flows accounts for 16% of global electricity and is expected to increase annually. It is cost-effective, renewable, and emits very little carbon dioxide. Despite its benefits, the construction of dams poses challenges to ecosystems and communities, presenting geological concerns, although it remains a dependable and environmentally friendly energy source.

The study [25] critically examines the integration of hydroelectricity into power systems, discussing wave pumps and storage systems. It addresses the environment, high upfront costs, and oil price uncertainty as challenges and discusses recent developments in government technology and institutional issues, especially in countries, and industries. For instance,

Turkey's hydropower plants (HPPs) [26], which contribute significantly to the country's electricity generation. About 25% of Turkey's electricity comes from HPP. Turkey has huge untapped hydropower potential in 26 reservoirs.

Geothermal energy from Earth's warming accounts for 0.3% of the world's electricity [27, 28]. Developing countries, including the Philippines and Kenya, rely on it. There are also nonelectric features such as heating and hydroponics. Geothermal is environmentally friendly, ranging from 2 to 10 US cents/kWh. The structure of hot and dry rocks suggests a possibility. The potential of enhanced geothermal systems (EGS) as a primary source of electricity in the USA is significant. In the year 2050, the increasing energy demands and environmental concerns underscore the importance of focusing on EGS as they offer a sustainable renewable energy solution for long-term preservation. Biomass currently accounts for 14% of global energy consumption, equivalent to approximately 25 million barrels of oil per day. It is important in developing countries (35% of their potential) and important in some industrialized countries, such as the USA (4%) and Sweden (14%) [29, 30]. The challenges include sustainable and cost-effective expansion. Biomass energy supply systems can be upgraded for electricity, liquid fuel, and gas. Various case studies show success and consider economic, social, and environmental factors. The *Biomass Energy Data Book* from Oak Ridge National Laboratory [31] provides comprehensive information on the biomass industry, including feedstocks, biofuels (ethanol, biodiesel, and bio-oil), biopower, biorefineries, and feedstock.

To limit warming to 1.5°C, 60% of oil, 60% of fossil methane gas, and 90% of coal must remain nonextractable by 2050. Global production must decline by 3% per year by 2050. These unclaimed resources vary from region to region, with the Middle East and Canada experiencing higher percentages. By 2050, the use of fossil fuels will further reduce these figures. The ability to achieve net zero emissions by 2050 is uncertain due to industry and technological constraints [32]. This depletion of fossil fuel reserves and the need to transition to sustainable carbon-free energy to combat climate change highlights the importance of new technologies and investments [33]. A hybrid energy mix including hydrogen, biofuels, and renewable energy is recommended.

The text traces the rise of fossil fuels with an emphasis on coal, from the use of "sea coal" in the 13th century to the Industrial Revolution and beyond, coal production doubled every 17 years from 1864 to 1914, indicating a rapid expansion in the use of fossil fuels [34]. Radioecology encompasses the environmental and radiative aspects of radioactivity in the environment. This includes radionuclide behavior, transport prediction, and the biological effects of ionizing radiation. This book is a comprehensive guide to radio biology [35]. Nuclear power, which accounts for 13% of global electricity, produces steady energy. Material challenges include corrosion, pressure breakdown, neutron weakness, and fuel system reliability. Radiation-induced effects and corrosion challenges need to be addressed in emerging fission–fusion systems [36].

The urgency of addressing climate change has led to a resurgence of interest in nuclear power. The work [37] presented nuclear fuel cycle dynamics and greenhouse gas emissions, focusing on the reasons for the shift. The first type of reactors, light water reactors and heavy water reactors require an energy flow of 0.1–0.3 kWh or per kWh. This results in strong greenhouse gases of 10–130 g CO_2-e/kWhel, less than fossil fuels (usually 600–1200 g CO_2-e/kWhel), but in air and water energy (about 15–25 g CO_2), more than -e/kWhel) and like solar energy (about 90 g CO_2-e/kWhel).

2.1.2 THE FUTURE THREATS TO ENERGY

The future of energy presents several significant challenges and threats that must be urgently managed. One of the biggest concerns is climate change. The continued reliance on fossil fuels and the resulting greenhouse gases seriously threatens our planet. The need to shift to cleaner and more sustainable energy sources is essential to mitigate the devastating consequences of climate change. Energy security is another important issue. Geopolitical tensions and the possibility of supply disruptions and increasing global energy demand leave our energy system vulnerable. The development of flexible and diverse energy is critical to meeting these challenges. Wasting limited resources is an ongoing concern. Eventually, fossil fuels like oil and natural gas will become scarce and expensive. To ensure stable and affordable energy, we need to invest in new energy sources and technologies. Finally, the outdated infrastructure of many energy systems around the world needs to be modernized for efficiency and reliability. In order to achieve a sustainable and prosperous energy future, it is imperative to tackle these challenges by fostering innovation, implementing policy reforms, and enhancing international cooperation.

This work integrated multiple energy sources in our IoT and Machine learning-based smart grid system (SGS). Architecture is presented, which takes appropriate decisions to maximize the use of renewable energy sources by predicting the next 10 days use of electricity in a particular home, and the next 10 days weather conditions. Section 2.2 of this chapter explains the IoT and machine learning methodology proposed. Section 2.3 contains the result analysis of the whole work, while Section 2.4 provides the concluding remarks.

2.2 PROPOSED METHODOLOGY

In this section, the architecture of the proposed system is designed, and the grid system technology is also presented. The prime function of the grid system is to predict the electric consumption of a particular home for the next ten days and take the weather conditions of the next ten days from the Internet. It also analyzes the current generated from solar and wind energy, schedules all the work and its energy source, and tries to increase the use of renewable energy sources. The following subsection explains the implemented architectures and machine learning models.

2.2.1 IoT ARCHITECTURE FOR GRID TECHNOLOGY

In the IoT architecture, the SGS takes three energy sources. One is the distributed electricity source (first source), solar energy source (second source), and wind energy source (third source), as shown in Figure 2.1.

The first source is connected to the automated control unit through an energy meter. This user is only used for the backup process. Second sources are connected to the smart grid through the inverter, as well as wind energy sources connected to the smart grid using the inverter. Through an inventor, the energy is filtered and converted to the AC and DC. The SG (abstract circuit diagram is shown in Figure 2.2) takes the energy sources as input.

SG is the central node of architecture. It connects with the Internet to store the data in the cloud and provides the interface to communicate with the end user. It also connects with the Internet to predict the next 10 days weather prediction. The SGS utilizes its monitoring capabilities to connect to the local Wi-Fi network for controlling electrical appliances. It dynamically generates a list of the appliances in its internal memory, allowing users to schedule and optimize the use of renewable energy sources efficiently. The subsequent paragraph will delve into the operational features of the device.

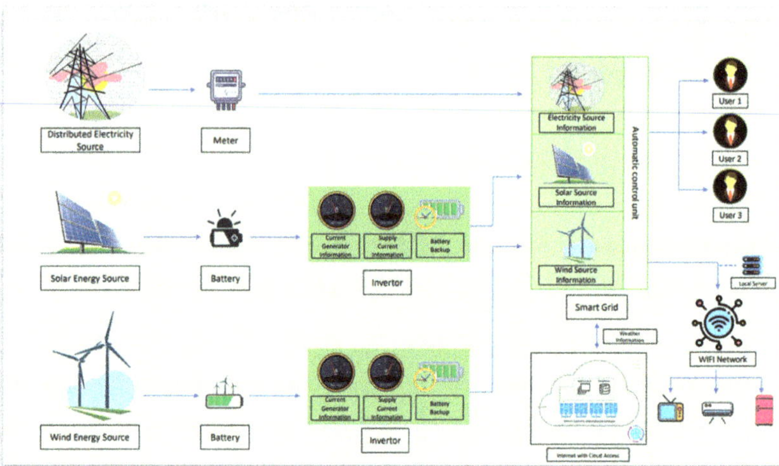

FIGURE 2.1 IoT Architecture for smart grid.

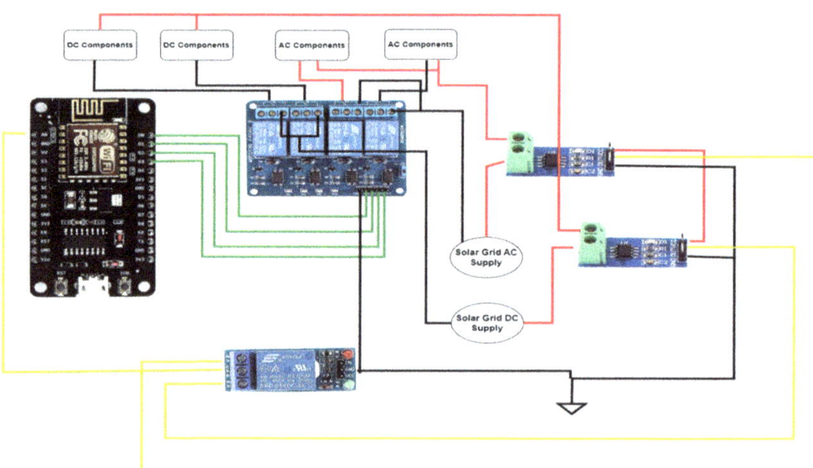

FIGURE 2.2 Solar grid architecture.

In this SGS system, MpdeMCU is used as the processing unit. Using NodeMCU can connect Wi-Fi network, with eight digital pins and one analogue pin. For demonstration purposes, we used the ACS712 Hall effect sensor for sensing the current. The experiment platform and computer communication have been given in Figure 2.3. The yellow line of Figure 2.2 is used to read data from the sensor, the green line denotes the trigger of the home appliance for on and off state, and the red and black line denote the voltage supply for each module.

FIGURE 2.3 Experiment platform.

2.2.2 ANALYTICAL MODEL

To prepare our analytical model, we take the six-year electricity consumption record of a home where only one person lives. The information was collected from Orissa Power Transmission Corporation Ltd. In this information, the attributes are "CIR," "DIVN", "SUBDIV," "SECTION," "ROUTE," "CONS_ACC," "CA_NO," "CONS_CODE," "CON_NAME," "CON_ADD1," "CON_ADD2," "CUR_STAT," "BILL_MON," "CAT_CODE," "CON_LOAD," "BIL_DATE," "BILBSS," "CON_LOAD.1," "MTR_COND," "PRV_READ," "PRS_READ," "UNIT_BILLED," "CUR_EC," "CUR_ED," "CUR_MR," "CUR_DC," "CUR_DPS," "TOTAL_BILL," "RB_AMT," "RB_DATE," "AMT_PAID," "COL_DATE," "BILL_MON.1," "CR_AMT," "DB_AMT," "NET_ARR." From all attributes, we used "CON_NAME," which denotes the consumer name, and "CON_ADD1" used for consumer address, to identify the consumer. For the machine learning model, we used the attributes "BILL_MON," which denotes the bill paid month like (JAN, FEB, …), "BIL_DATE" shows the date on which the bill is paid, "UNIT_BILLED" the billed unit amount for the particular month, and "TOTAL_BILL" is amount paid for this month. The basic process of data science is shown in Figure 2.4.

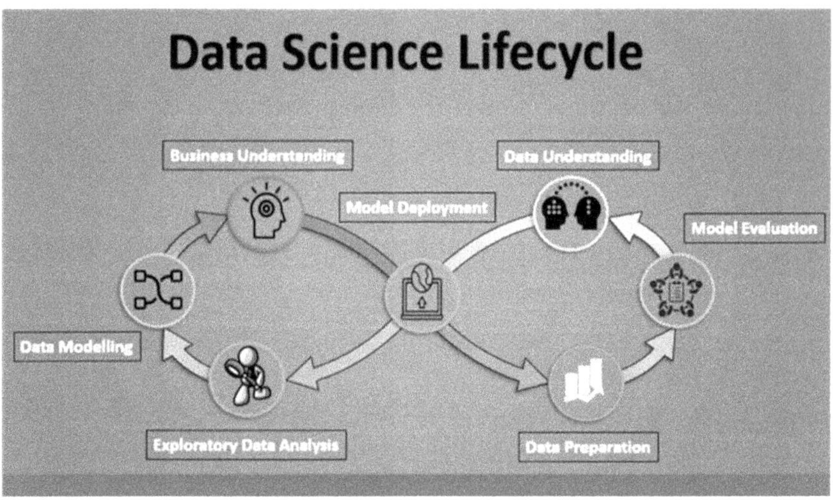

FIGURE 2.4 Life cycle of data science.

In the presented work, we used linear regression as a machine learning model. In our model, we put the month as input and find the predicted billing amount. We find the accuracy level up to 78.96. The whole process is explained below.

- **Task Description:**
  - Task 1: Predicting monthly electricity consumption based on the provided month and consumer name.
  - Task 2: Analytically forecasting electricity consumption for a specific number of days starting from a given day in a month.

- **Data collection:**
  - Data Source Selection: Collate historical electricity consumption data from your home for six years. This dataset will serve as the primary data source for model development.
  - Data Preparation: Ensure the dataset includes detailed records of monthly consumption, including the month and corresponding consumption values.

- **Data Exploration and Analysis:**
  - Explore the data to understand its characteristics.
  - Visualize data distributions and relationships between variables.
  - Identify patterns, anomalies, or missing values.

- **Data Cleaning:**
  - Handling Missing Values: Identify and address any missing or null values in the dataset, particularly in the electricity consumption and date columns. This can be done through methods like imputation, removal of incomplete records, or interpolation, ensuring that the dataset is complete and reliable.
  - Date Format Standardization: Standardize date formats to ensure consistency across the dataset. Some dates may not be in the required format, and you'll need to transform them into a uniform structure. This ensures that the predictive and analytical models can effectively interpret the temporal data.
  - Outlier Detection and Treatment: Check for outliers in electricity consumption values, as they might affect model performance. Decide on an approach for handling outliers, which could involve removal, transformation, or capping.
  - Data Validation: Validate the dataset for data integrity, consistency, and correctness. Ensure that all records are relevant to the problem statement and that there are no duplicate entries.

  - Data Normalization: Normalize the data, if necessary, particularly if you have varying units or scales. This ensures that different features are on a similar scale, aiding model training.
  - Feature Engineering: If applicable, engineer additional features that may enhance model performance, such as creating time-based features (e.g., day of the week) that can help capture temporal patterns in electricity consumption.

- **Data Splitting:**
  - Split the data into training, validation, and test sets.
  - Ensure that the data split is representative and random.

- **Model Selection:**
  - Choose appropriate machine learning algorithms.
  - Consider the problem type (classification, regression, clustering, etc.).
  - Here we used a linear regression model.

- **Model Building:**
  - Develop your machine learning model(s).
  - Fine-tune hyperparameters, such as learning rates, regularization, and architecture.
  - Train the model on the training data.

- **Model Evaluation:**
  - Evaluate the model's performance on the validation set.
  - Utilize appropriate evaluation metrics (accuracy, F1-score, RMSE, etc.).
  - Make necessary adjustments to the model based on results.

- **Model Optimization:**
  - Optimize the model for better performance.
  - Consider techniques like hyperparameter tuning and regularization.

- **Model Testing:**
  - Assess the final model on the test data.
  - Ensure the model generalizes well to unseen data.

- **Results Interpretation:**
  - Analyze the model's predictions and understand its behavior.
  - Communicate insights gained from the model.

- **Model Deployment:**
  - Deploy the model to a production environment if it meets requirements.
  - Set up monitoring for model performance.

12.3 ANALYSIS AND RESULT

After successfully implementing our model, we found the following analytics, and the predicted result is explained below.

Analytics:

- Con
- sumer Account Number (CONS_ACC):
  - Unique identifier for each consumer.
  - Links consumers to their consumption and billing history.

- Billing Period Information:
  - BILL_MON tracks billing month.
  - BIL_DATE records the exact bill generation date each month.

  Together, they provide a timeline of billing periods for each consumer.

- Contracted *versus* Recorded Load:
  - CON_LOAD shows sanctioned load in KW as per consumer contract.
  - CON_LOAD.1 shows the actual load recorded by a meter each month.
  - Comparison shows if consumer exceeded their sanctioned load for the billing period.

- Energy Consumption:
  - PRV_READ and PRS_READ record meter readings at the start and end of each billing period.
  - UNIT_BILLED calculates units consumed based on meter reading difference.

   - Tracks monthly energy consumption for each consumer.

- Billing Amounts:
   - CUR_EC, CUR_ED, CUR_MR, CUR_DC, and CUR_DPS provide per-unit charges for energy, duties, meter rent etc.
   - TOTAL_BILL calculates the monthly billing amount by summing all charge components.
   - It shows how consumption translates into billing amount each month.

- Adjustments to Billing Amount:
   - Rb_AMT tracks any rebate provided to the consumer for the billing period.
   - CR_AMT tracks any credit adjustments for past dues/ overpayments.
   - These are adjusted against TOTAL_BILL to derive NET_ARR (Net amount payable)

- Payment History:
   - AMT_PAID records actual payments made against each month's bill.
   - Tracks payment status month-wise for each consumer.

- Meter Reading Type:
   - MTR_COND indicates if billing is based on Actual or Provisional meter readings.
   - Provisional means reading not taken for that billing period.

2.3.1 RESULTS

We successfully predicted the following result, which is displayed using a graph that is Figure 2.5. Graph 1 explains the predicted electricity consumed in every month. Graph 2 shows the predicted bill amount for each month. Graph 3 shows the total bill *versus* the total unit of electricity consumed. Graph 4 explains the monthly billing amount over time in a blue line. The monthly payment received in an orange line, the total units consumed per month in the green line, and the average energy charge per month in the red line.

The output of weather prediction is given in Figure 2.6.

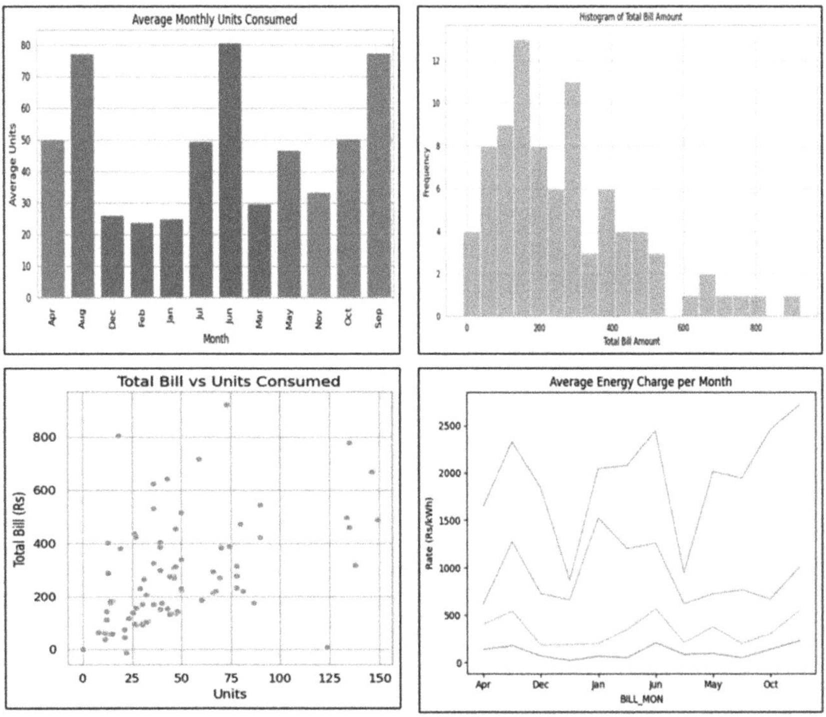

FIGURE 2.5 We found out the results from the analysis.

```
Enter the city name: gunupur
Enter the start date (YYYY-MM-DD): 2003-02-12
Weather forecast for gunupur for the next 10 days from 2003-02-12:

Date: 2023-09-27, Weather: Thunderstorm with rain, Temperature: 26.9°C
Date: 2023-09-28, Weather: Thunderstorm with rain, Temperature: 27°C
Date: 2023-09-29, Weather: Thunderstorm with rain, Temperature: 27.1°C
Date: 2023-09-30, Weather: Moderate rain, Temperature: 25.9°C
Date: 2023-10-01, Weather: Heavy rain, Temperature: 25.9°C
Date: 2023-10-02, Weather: Heavy rain, Temperature: 25.1°C
Date: 2023-10-03, Weather: Thunderstorm with heavy rain, Temperature: 26.2°C
Date: 2023-10-04, Weather: Scattered clouds, Temperature: 26.2°C
Date: 2023-10-05, Weather: Scattered clouds, Temperature: 26°C
Date: 2023-10-06, Weather: Scattered clouds, Temperature: 26.1°C
```

FIGURE 2.6 Result of the next 10 days weather prediction.

Using the above implementation, the proposed SGS can merge the energy source and can make intelligent decisions to make a home appliance used scheduler for maximum use of renewable energy and less use of distributed energy sources, which can only be used for the backup process.

2.4 CONCLUSION

SGS offers a smart solution for mitigating potential risks in the energy industry. By implementing this system, customers can optimize their utilization of renewable energy sources while reducing reliance on a decentralized approach, resulting in a more cost-efficient solution. It also minimizes the environmental-related issue created by nonrenewable energy sources. The accuracy level of the prediction is 78.96, indicating the intelligence of SGS.

KEYWORDS

- **solar energy**
- **grid technology**
- **wind energy**
- **sustainable energy**

REFERENCES

1. Jun, E., Kim, W., & Chang, S. H. The analysis of security cost for different energy sources. *Applied Energy* **2009**, *86*(10), 1894–1901.
2. Forsberg, C. W. Sustainability by combining nuclear, fossil, and renewable energy sources. *Progress in Nuclear Energy* **2009**, *51*(1), 192–200.
3. Dombaycı, Ö. A., Gölcü, M., & Pancar, Y. Optimization of insulation thickness for external walls using different energy-sources. *Applied Energy* **2006**, *83*(9), 921–928.
4. Mohammadi, A., & Mehrpooya, M. A comprehensive review on coupling different types of electrolyzer to renewable energy sources. *Energy* **2018**, *158,* 632–655.
5. Chien, F., Kamran, H. W., Albashar, G., & Iqbal, W. Dynamic planning, conversion, and management strategy of different renewable energy sources: A sustainable solution for severe energy crises in emerging economies. *International Journal of Hydrogen Energy* **2021**, *46*(11), 7745–7758.
6. Vega, A. M., Santamaria, F., & Rivas, E. Modeling for home electric energy management: A review. *Renewable and Sustainable Energy Reviews* **2015**, *52*, 948–959.

7. Mowad, M. A. E. L., Fathy, A., & Hafez, A. Smart home automated control system using android application and microcontroller. *International Journal of Scientific & Engineering Research* **2014**, *5*(5), 935–939.
8. Zhou, K., Yang, S., & Shao, Z. Energy internet: The business perspective. *Applied Energy* **2016**, *178*, 212–222.
9. Al-Ali, A. R., Zualkernan, I. A., Rashid, M., Gupta, R., & AliKarar, M. A smart home energy management system using IoT and big data analytics approach. *IEEE Transactions on Consumer Electronics* **2017**, *63*(4), 426–434.
10. Ahmad, T., & Zhang, D. Using the internet of things in smart energy systems and networks. *Sustainable Cities and Society* **2021**, *68*, 102783.
11. Tom, R. J., & Sankaranarayanan, S. IoT based SCADA integrated with Fog for power distribution automation. In: *Proceeding of 12th Iberian Conference on Information Systems and Technologies (CISTI)*, **2017**, pp. 1–4.
12. Al-Turjman, F., Altrjman, C., Din, S., & Paul, A. Energy monitoring in IoT-based ad hoc networks: An overview. *Computers & Electrical Engineering* **2019**, *76*, 133–142.
13. Saleem, Y., Crespi, N., Rehmani, M. H., & Copeland, R. Internet of things-aided smart grid: technologies, architectures, applications, prototypes, and future research directions. *IEEE Access* **2019**, *7*, 62962–63003.
14. Omnes, N., Bouillon, M., Fromentoux, G., & Le Grand, O. A programmable and virtualized network & IT infrastructure for the internet of things: How can NFV & SDN help for facing the upcoming challenges. In: *Proceeding of 18th International Conference on Intelligence in Next Generation Networks*, **2015**, pp. 64–69.
15. Sundhari, R. M., & Jaikumar, K. IoT assisted Hierarchical Computation Strategic Making (HCSM) and Dynamic Stochastic Optimization Technique (DSOT) for energy optimization in wireless sensor networks for smart city monitoring. *Computer Communications* **2020**, *150*, 226–234.
16. Devi, M., Dhaya, R., Kanthavel, R., Algarni, F., & Dixikha, P. Data science for Internet of Things (IoT). In: *Second International Conference on Computer Networks and Communication Technologies (ICCNCT)*, **2019**, pp. 60–70.
17. Marjani, M., Nasaruddin, F., Gani, A., Karim, A., Hashem, I. A. T., Siddiqa, A., & Yaqoob, I. Big IoT data analytics: Architecture, opportunities, and open research challenges. *IEEE Access* **2017,** *5,* 5247–5261.
18. Adi, E., Anwar, A., Baig, Z., & Zeadally, S. Machine learning and data analytics for the IoT. *Neural Computing and Applications* **2020**, *32*, 16205–16233.
19. Marculescu, R., Marculescu, D., & Ogras, U. Edge AI: Systems design and ML for IoT data analytics. In: *Proceedings of the 26th ACM SIGKDD International Conference on Knowledge Discovery & Data Mining*, **2020**, pp. 3565–3566.
20. Gong, J., Li, C., & Wasielewski, M. R. Advances in solar energy conversion. *Chemical Society Reviews* **2019**, *48*(7), 1862–1864.
21. Mazria, E. *Passive Solar Energy Book*, Rodale, PA, USA **1979**, p. 448.
22. Kannan, N., & Vakeesan, D. Solar energy for future world: A review. *Renewable and Sustainable Energy Reviews* **2016**, *62*, 1092–1105.
23. Tiwari, R., & Babu, N. R. Recent developments of control strategies for wind energy conversion system. *Renewable and Sustainable Energy Reviews* **2016**, *66*, 268–285.
24. Bagher, A. M., Vahid, M., Mohsen, M., & Parvin, D. Hydroelectric energy advantages and disadvantages. *American Journal of Energy Science* **2015**, 2(2*)*, 17–20.

25. Sims, G. P. Hydroelectric energy. *Energy Policy* **1991**, *19*(8), 776–786.
26. Ozturk, R., & Kincay, R. Potential of hydroelectric energy. *Energy Sources* **2004**, *26*(12), 1141–1156.
27. Barbier, E. Geothermal energy technology and current status: An overview. *Renewable and Sustainable Energy Reviews* **2002**, *6*(1–2), 3–65.
28. Fridleifsson, I. B. Geothermal energy for the benefit of the people. *Renewable and Sustainable Energy Reviews* **2001**, *5*(3), 299–312.
29. Hall, D. O. Biomass energy. *Energy Policy* **1991**, *19*(8), 711–737.
30. Field, C. B., Campbell, J. E., & Lobell, D. B. Biomass energy: The scale of the potential resource. *Trends in Ecology & Evolution* **2008**, *23*(2), 65–72.
31. Boundy, R. G., & Davis, S. C. *Biomass Energy Data Book: Edition 3 (No. ORNL/TM-2011/43)*. Oak Ridge National Lab. (ORNL): Oak Ridge, TN, **2010**.
32. Welsby, D., Price, J., Pye, S., & Ekins, P. Unextractable fossil fuels in a 1.5°C world. *Nature* **2021**, *597*(7875), 230–234.
33. Abas, N., Kalair, A., & Khan, N. Review of fossil fuels and future energy technologies. *Futures* **2015**, *69,* 31–49.
34. Hubbert, M. K. Energy from fossil fuels. *Science* **1949**, *109*(2823), 103–109.
35. Whicker, F. W., & Schultz, V. Radioecology: Nuclear energy and the environment, **1982**.
36. Zinkle, S. J., & Was, G. S. Materials challenges in nuclear energy. *Acta Materialia* **2013**, *61*(3), 735–758.
37. Lenzen, M. Life cycle energy and greenhouse gas emissions of nuclear energy: A review. *Energy Conversion and Management* **2008**, *49*(8), 2178–2199.

CHAPTER 3

The Use of the Internet of Things for Active Noise Control Technology

SUMAN TURPATI[1,*], VALLI BHASHA[2], and AJAY ROY[3]

[1]*Department of ECE, Rajeev Gandhi Memorial College of Engineering and Technology, Nandyal, Andhra Pradesh, India*

[2]*Department of Electronics and Communication Engineering, Sri Venkateswara College of Engineering & Technology, Chittoor, Andhra Pradesh, India*

[3]*Department of Electrical Engineering, College of Engineering, University of Bisha, Bisha,Kingdom of Saudi Arabia*

*Corresponding author

ABSTRACT

Active noise cancellation (ANC) is a well-established field that involves the cancellation of noise in the surrounding environment by the generation of anti-noise signals in close proximity to the human ears. This technology is often used in products such as noise cancellation headphones. This research article introduces the integration of wireless communication and acoustics in order to include the Internet of Things (IoT) into active noise cancellation technology. The fundamental concept involves the deployment of IoT device inside a given location, which is designed to capture and process ambient sounds. Subsequently, the device transmits the recorded sound data through its wireless radio connectivity. The use of ANC technology has significant implications not just within the field of IoT, but also across several other domains such as system identification, noise suppression, communication, and signal management, among others.

IoT Potential for Green Energy Solutions. Abhishek Kumar, Ajay Roy, & Shamimul Qamar (Eds.)

DOI: 10.1201/9781779643599-3

Due to the higher velocity of wireless signals compared to sound waves, the auditory apparatus in our ear detects the sound prior to its physical arrival. This provides a view into the future, referred to as look ahead, and is of significant importance for the real-time mitigation of noise, particularly for unpredictable and wide-band auditory stimuli such as music and speech. The aim of this study is to explore the practical uses of the ANC technique and assess the effectiveness of an adaptive algorithm. Additionally, this study presents the use cases of three noise cancelling filters in the context of the IoT. This chapter will conclude with a comprehensive performance evaluation of several ANC approaches, including recursive least square, variable step-size least mean square (LMS), leaky LMS, normalized least mean square, and LMS, based on the data obtained through simulations.

3.1 INTRODUCTION

The Internet of Things (IoT) has gained significant attention and is increasingly being integrated into everyday life, as a result of advancements in science and technology. The emergence of a growing industry and the nation's prosperity, as well as the enhancement of people's quality of life, are symbolized by the growth of the IoT. Moreover, communication technology is developing at a faster rate at the same time. On the other hand, noise and distortion may seriously impair wireless communication and data exchange across wireless channels as technology can only survive with dependability and high customer satisfaction. Because of this, we were inspired to consider adaptive noise-cancelling methods that would be useful in this kind of circumstance.

Numerous methodologies have been proposed to mitigate or regulate acoustic noise, including passive and active noise management, which are two distinct ways. Passive noise control (PNC) refers to the conventional approach used for reducing acoustic noise levels by using barriers, enclosures, and silencers. In this case, the absorption material is used to convert the acoustic energy into heat energy. High-frequency applications may be effectively controlled using the PNC technique. Passive approaches, on the other hand, are not very good at managing low frequencies and are huge and expensive. Because of this, the passive approach is not feasible for a wide range of low-frequency applications. To solve these drawbacks, the idea of active noise control (ANC) has been studied [1–5].

The active noise reduction technology was first created in 1936 by Paul Lueg, US Patent. Over the last several decades, with the advent of low-cost digital signal processing (DSP) and optimized control algorithms to attenuate low-frequency noise, researchers have become more interested in ANC. This method demonstrated a strong desire to solve the shortcomings of PNC suppression. Particularly for frequencies below 1000 Hz, the approach might be appealing for reducing noise in a compact container and has several benefits over passive methods. Compared to passive approaches, it offers certain potential benefits in terms of cost, size, and quick gains in noise reduction.

3.2 ACTIVE NOISE CONTROL

The ANC [6–10] is a system that operates by the creation of antinoise signals, using an electroacoustic mechanism. The performance of the system is based on the idea of destructive interference, as shown in Figure 3.1. The concept behind the ANC system was to produce a control signal, known as antinoise, which had the same amplitude but opposite phase as the primary noise. The primary noise and control sounds are combined, resulting in a residual noise that has a reduced magnitude compared to the original noise. The effectiveness of an ANC system in reducing noise levels is dependent upon the precision with which control sound intensity is generated. The ANC system has shown to be proficient at controlling low-frequency applications, offering advantages in terms of affordability, compactness, and simplicity. Consequently, it has garnered more recognition in comparison to PNC systems. Figure 3.2 illustrates the fundamental ANC, which encompasses the noise source, control noise, and residual noise components.

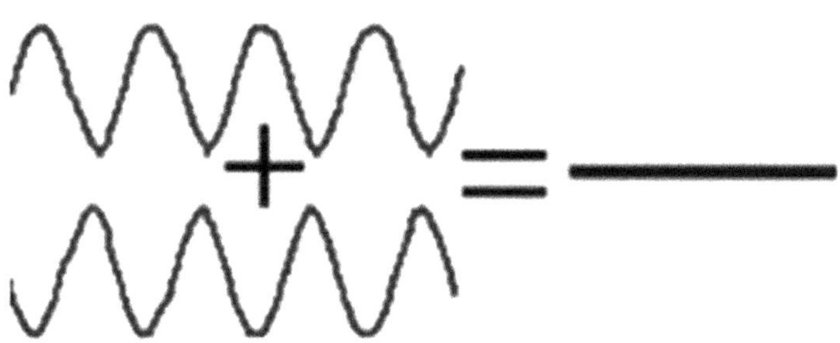

FIGURE 3.1 The idea of destructive interference.

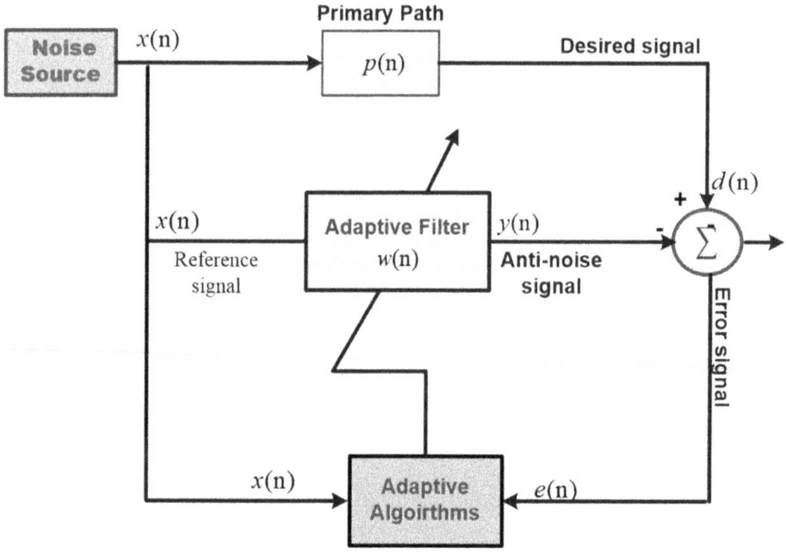

FIGURE 3.2 Block diagram of active noise control.

The concept of a basic ANC structure, as seen in Figure 3.2, necessitates the incorporation of adaptive principles to successfully reduce noise in environments where acoustic sources exhibit non-stationary behavior. The functioning of the system is primarily carried out *via* the use of four key components. The components involved in this system are as follows: (1) sensors, which serve as transducers, are responsible for converting acoustic signals into their corresponding electrical form. These electrical signals are then utilized to provide the controller with both a reference signal and a residual signal; (2) actuators, specifically loudspeakers, are employed to generate control signals. These control signals are required to be in the form of acoustic signals to minimize noise; (3) the plant refers to the physical system that is being controlled, such as a DUCT; (4) the controller acts as the central component of the system. It utilizes adaptive algorithms to control the noise by generating appropriate control signals based on the noise obtained from the sensors. An adaptive filter is primarily defined by its ability to independently modify the weight coefficient following the surrounding variations. In general, the adaptive filter maintains the same structure and solely adjusts its parameters. Adaptive filters possess the ability to adapt their filter coefficients based on the changing statistical features of input signals, without requiring prior knowledge of the input and output statistical properties. This is achieved by the use of a particular algorithm.

3.2.1 THE ADAPTIVE ALGORITHM COMPRISES TWO ESSENTIAL PROCESSES

3.2.1.1 FILTERING PROCESS

The filtering process entails the computation of the transverse filter response based on the tap inputs, followed by a comparison of this filter response with the intended signal. This comparison yields a residual error.

3.2.2 ADAPTIVE PROCESS

The adaptive process involves the automated modification of filter coefficients (weights) in response to the reference input and error estimates. To ensure the filter remains in its ideal filtering state, it is necessary to monitor and continually alter the coefficients and related characteristics at any given time throughout the filtering process. Consequently, the adaptive filter has a high level of adaptability.

3.2.2 THREE PRIMARY ELEMENTS MAKE UP THE ANC SYSTEM

3.2.2.1 ADAPTIVE FILTER

The adaptive filter is a kind of digital filter that utilizes the input signal to figure out and produce the filter's output. Based on the inherent properties of the transfer function of a filter, it is possible to categorize it into two distinct types: linear filters and nonlinear filters. The designer is responsible for determining the structure of this component, and its parameters are modified based on the algorithm's outcome.

3.2.2.2 RESIDUAL PERFORMANCE CRITERION

The performance requirements for the filter are often assessed based on its ability to provide results that align with the anticipated response.

The average form of square error is often used for practical system design and mathematical processing simplicity.

3.2.2.3 ADAPTIVE ALGORITHM

The adaptive algorithm is responsible for adjusting the filter settings to enhance the filter's performance. The performance index, input signal, and anticipated response should all be taken into consideration while choosing the adaptive approach.

The block diagram of the ANC system is shown in Figure 3.2. In this scenario, a reference signal is acquired from either sources of noise or a sensor with an unknown frequency. This signal is then sent over a primary channel, resulting in the generation of a desired signal. In Figure 3.2, the adaptive filter is linked in parallel to the primary channel. As a result, it gets stimulated by the same reference signal and creates an antinoise signal, which is then subtracted from the desired signal, resulting in the generation of a residual (error) signal. The residual signal that has been acquired is gradually reduced by modifying the filter weights *via* the use of various adaptive algorithms.

3.3 ADAPTIVE ALGORITHMS

3.3.1 LEAST MEAN SQUARE ALGORITHM

The least mean square (LMS) method is a collection of stochastic linear-gradient algorithms designed to determine the optimal filter weights (coefficients) for effectively minimizing the instantaneous error signal. Initially, it offers a way to track statistical signals and variations. Secondly, it eliminates the necessity for obtaining exact knowledge of statistical signals. Lastly, the LMS method considers the absence of error-signal information in diverse applications [3, 4].

The LMS algorithm's weight update equation is as follows:

$$w(n+1) = (-\nabla J)\frac{\mu}{2} + w(n) \tag{3.1}$$

where J is the cost function, μ is the step-size factor, and $w(n+1)$ denotes the updating filter weights.

This algorithm's cost function is written as

$$J = e^2(n) \tag{3.2}$$

The error l $e(n)$ is

$$e(n) = d(n) - y(n) \qquad = d(n) - x(n)\, w^T(n) \tag{3.3}$$

By using the abovementioned Equation 3.3, the derivative of the cost function concerning $w(n)$ may be expressed as

$$\nabla J = \nabla e^2(n) = -2e(n)\, x(n) \tag{3.4}$$

Using Equation 4, the weight updating of the LMS algorithm is

$$w(n + 1) = w(n) + \mu e(n)\, x\,(n) \tag{3.5}$$

The condition must be followed while considering the step size value to prevent the divergence issue.

$$0 < \mu < \frac{2}{LP_x} \tag{3.6}$$

where P_x is the auto-correlation matrix of the filter and L is its order.

3.3.2 NORMALIZED LMS ALGORITHM

The performance of the LMS method's metric parameters, including convergence rate and stability, is contingent upon the autocorrelation matrix of the input vector. However, the algorithm encountered difficulties in terms of gradient noise scaling as a result of the substantial size of the input vector. To address such difficulties, a normalized LMS method was created. The adaptation of the LMS algorithm presents a challenge in selecting an appropriate value for the step size μ, mostly owing to its huge magnitude. The NLMS algorithm has included the notion of variable step size to address the issue of divergence in the parameter μ [2].

The adaptive step size is defined as

$$\mu(n) = \frac{\mu}{\delta + \|x(n)\|^2} \tag{3.7}$$

where δ is a small constant and $\|\cdot\|$ represents a Euclidian form.

The weight adaptation of the normalized LMS method is mathematically represented as

$$w(n+1) = \mu(n)\, e(n)\, x(n) + w(n) \tag{3.8}$$

3.3.3 LEAKY LMS ALGORITHM

In practical applications, the use of digital filters serves to effectively mitigate issues such as quantization inaccuracy. The LMS algorithms use computing processes such as addition and multiplication to update the weights of the filters. In the context of practical approaches, namely fixed point board DSP, the previous computations may result in quantizing mistakes, which refers to the situation where the adaptive weights have the potential to diverge. To mitigate the issue of divergence, a "Leaky" method might be included in the weight updating process. The Leaky LMS method was proposed as a solution to address the limitations encountered in the conventional LMS technique [2].

Furthermore, the Leaky LMS's cost function comprises the sum of squared filter coefficient responses for the square error function; this may be stated mathematically as

$$J(n) = e^2(n) + \alpha \sum_{i=0}^{L-1} w_i^2(n) \tag{3.9}$$

where α represents the leakage factor is $0<\alpha<1$.

The following is the derivation of the cost function's gradient estimation:

$$\nabla J(n) = -2\, x(n)\, e(n) + 2\alpha\, w(n) \tag{3.10}$$

The way leaky LMS adjusts for weight is stated

$$w(n+1) = \mu e(n)\, x(n) + w(n)\,(1-\mu\alpha) \tag{3.11}$$

3.3.4 ARIABLE STEP-SIZE LMS ALGORITHM

One notable feature of the LMS method is that it incorporates the convergence factor, or step size, not only for updating the filter weights but also for generating the Excess Mean Square Error (EMSE). The efficacy of the convergence speed in the LMS algorithm is contingent upon the magnitude of the step size. The values exhibit a tiny EMSE and a slow convergence speed for the low step size, whereas large step size values lead to a fast convergence speed and EMSE. The resolution of this issue was achieved by the use of a variable step-size LMS method. The given text consists of two numbers, specifically 2 and 29. The value of each step is subject to variation based on the estimation of the distance to the LMS error (MSE) [2].

The weight adaption of the VSS-LMS algorithm is expressed as

$$w(n+1) = w(n) + U(n)\, e(n)\, x(n) \tag{3.12}$$

where $U(n)$ is the diagonal matrix of step size μ_{I} for individual weights at each time n

$$U(n) = \begin{bmatrix} \mu_0(n) & 0 & 0 \\ 0 & \mu_1(n) & 0 \\ 0 & 0 & \mu_{L-1} \end{bmatrix} \tag{3.13}$$

3.5 ANC APPLICATIONS IN THE INTERNET OF THINGS

When data transmission between several senders and recipients occurs inside a network, active noise reduction may be very important in the IoT. Different atmospheric variations combined with channel quality cause the signal to get distorted as it travels *via* an air interface. When undesired components are eliminated from the signal, these noise cancellation methods may assist in returning the signal to its original state. Adaptable system reconfiguration based on channel state and context is possible in cognitive radio networks. Adaptive noise control filters may enhance signaling performance by modifying their settings based on the present circumstances. Two categories exist for commonly used adaptive filters that use noise-cancelling methods.

3.5.1 SIMULATION RESULTS AND ANALYSIS

The present study used simulation experiments to evaluate and analyze the performance of conventional LMS algorithms and their variants in terms of noise reduction and convergence. Several methods are implemented and evaluated inside noise-cancellation systems, to limit the effects of different statistical and real-time noise environments. Table 3.1 contains the simulation settings.

TABLE 3.1 Simulation Parameters

S. No.	Algorithm	Step-Size (μ)	δ	λ (Leakage)	No. of Iterations
1	LMS	0.003	–	–	200
2	NLMS	0.012	5	–	200
3	LLMS	0.0005	–	0.85	200
4	VSS LMS	–	–	–	200

3.5.2 CASE STUDY

3.5.2.1 CASE 1: THE RANDOM NOISE SIGNAL HAS A VARIANCE OF 0.2

A zero-mean white noise with a variance of 0.2 is used as the noise source. The impacts of the LMS, NLMS, LLMS, and VSS-LMS algorithms on control signal measurement are shown in Figure 3.3a–d, respectively. As the NLMS has 72.5% noise suppression, it obtains the least error signal magnitude when compared to the LMS, LLMS, and VSS LMS algorithms.

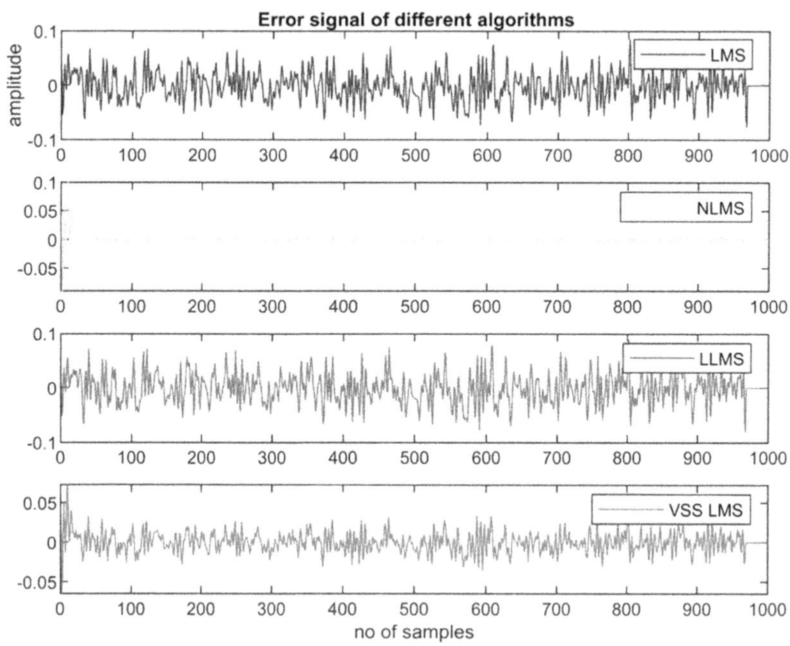

FIGURE 3.3 Employing various adaptive methods for noise reduction.

Figure 3.4 illustrates a performance comparison among the four methods. The convergence of LMS and LLMS algorithms is seen to commence after 180 iterations. Comparatively, VSS-LMS exhibits somewhat superior performance, with convergence beginning after 120 iterations. Notably, the NLMS method outperforms the three aforementioned algorithms, with convergence initiating after 75 iterations. The NLMS method has a favorable convergence speed, yet it also implies a greater level of computational complexity compared to other algorithms. In a situation when more rapid convergence is necessary, the LMS algorithm may not be the most suitable option. Instead, the NLMS algorithm would be a preferable alternative due to its faster convergence speed and improved efficiency.

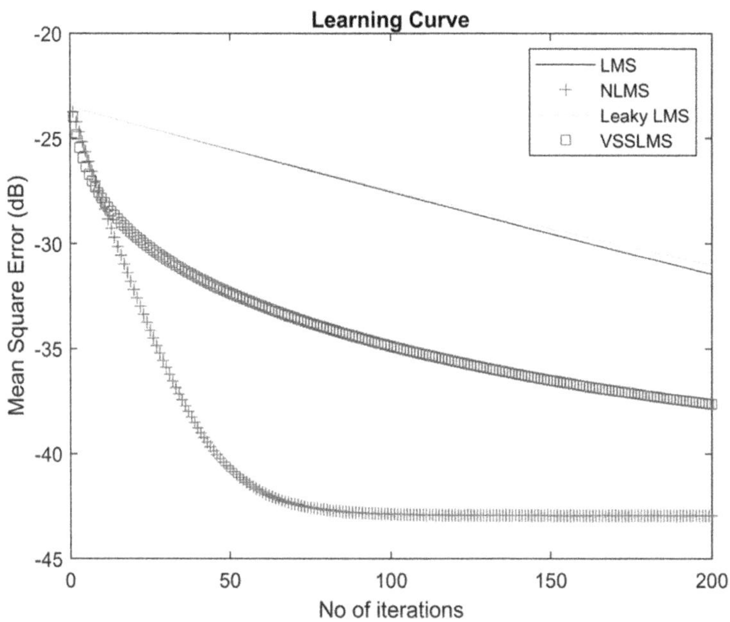

FIGURE 3.4 Compare the convergence performance of four different algorithms: LMS, NLMS, LLMS, and VSS-LMS

3.5.2.2 CASE 2: THE COLOR NOISE SIGNAL

The noise source used in this study is a color noise signal. Figure 3.5 illustrates the effects of the LMS, NLMS, LLMS, and VSS-LMS algorithms on the measurement of control signals. The NLMS method demonstrates a noise suppression capability of 68.85%, exhibiting superior performance in terms of minimizing error signal size when compared to the LMS, LLMS, and VSS LMS algorithms.

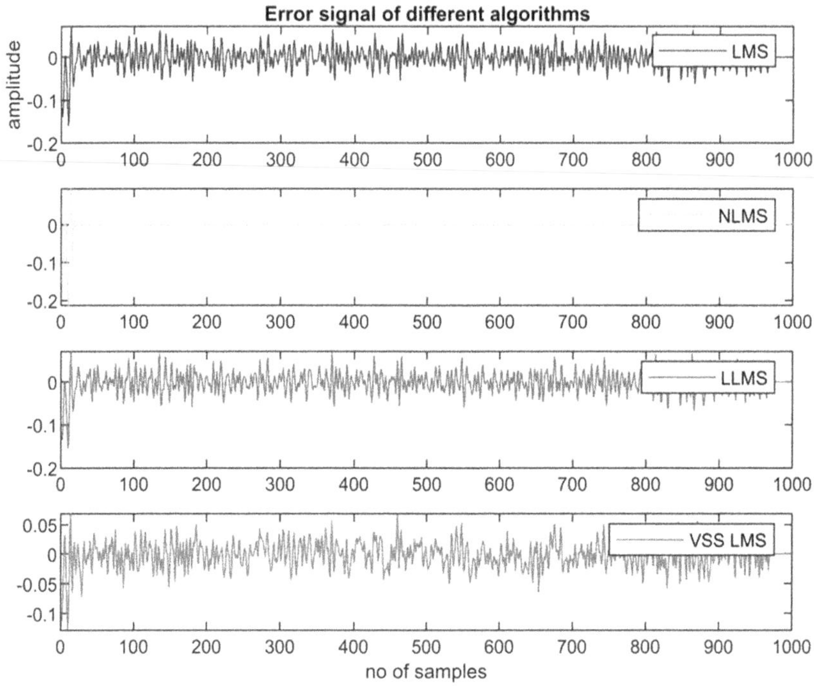

FIGURE 3.5 Using different adaptive techniques to reduce noise.

A performance comparison of the four techniques is shown in Figure 3.6. After 140 iterations, the convergence of the LMS and LLMS algorithms is seen to start. In comparison, VSS-LMS performs a little better, starting to converge after 90 iterations. NLMS performs better than the three previously stated algorithms, and convergence starts at iteration 50. Although the NLMS approach has a faster rate of convergence than other algorithms, it also suggests a higher degree of computational complexity. When a circumstance calls for a faster convergence, the LMS algorithm may not be the best choice. Rather, the NLMS algorithm would be a better choice because of its increased efficiency and quicker rate of convergence.

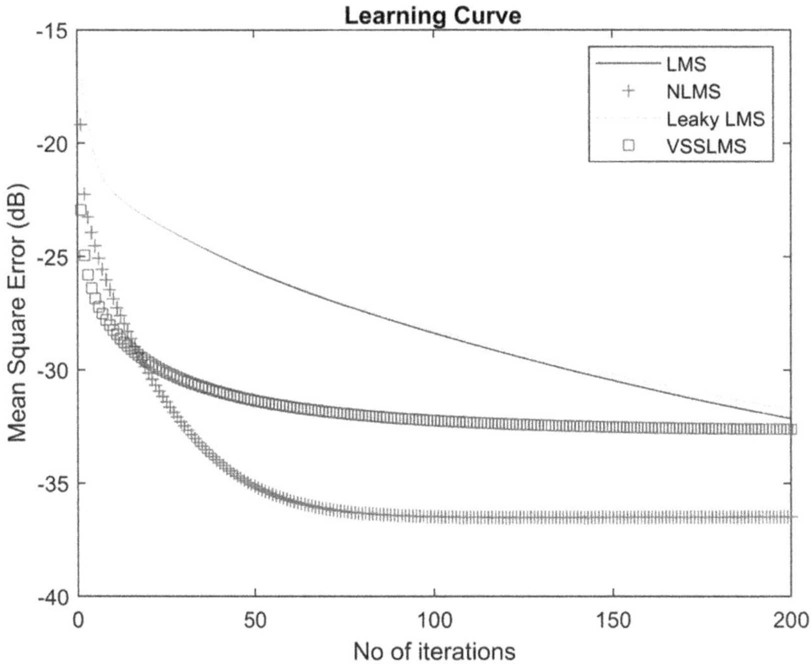

FIGURE 3.6 Compare the convergence performance of four different algorithms: LMS, NLMS, LLMS and VSS-LMS.

3.5.2.3 CASE 3:RECORDED FAN NOISE

Recorded fan noise was used as the noise source in this study. Figure 3.7 shows how the LMS, NLMS, LLMS, and VSS-LMS methods change the analysis of control signals. It is better at reducing error signal size than the LMS, LLMS, and VSS-LMS algorithms, with a noise reduction level of 68.85% demonstrated by the NLMS method. Illustration Figure 3.8 presents a comparative analysis of the performance of the four techniques. The LLMS and LMS algorithms are observed to begin convergent after 110 iterations. VSS-LMS performs marginally better, with convergence beginning after 80 iterations. Initial convergence of NLMS occurs at iteration 40, which is superior to the performance of the three algorithms mentioned earlier. The NLMS method suggests a greater degree of computational complexity, despite its quicker convergence rate in comparison to alternative algorithms. The LMS algorithm might not be the optimal selection when a quicker convergence is required. Considering its accelerated convergence and enhanced efficiency, the NLMS algorithm would be a more suitable alternative.

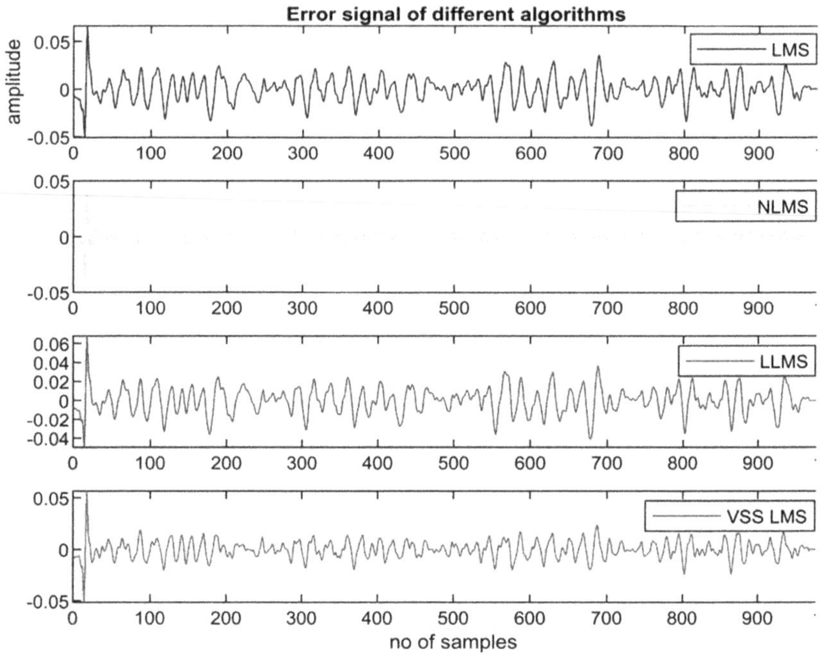

FIGURE 3.7 Noise reduction using a various adaptive algorithms.

3.6 CONCLUSION

The concept of adaptive filters and adaptable algorithms is examined in this work. A computational tool that illustrates the relationship between a filter's input and output signals is called an adaptive filter. The filter coefficients are adjusted by themselves based on an adaptive algorithm. Within the domain of the IoT, it may be beneficial to reduce noise and distortion that may arise from signals with high levels of interference. This mitigation process serves to improve the quality of services and boost overall dependability.

Numerous researchers and writers are engaged in the study and development of adaptive filtering algorithms with the aim of enhancing performance in the context of the IoT. Undoubtedly, the IoT represents the future of wireless communication. Therefore, the aim was to provide a thorough depiction of noise-cancelling adaptive filters. An evaluation was conducted on the effectiveness and converge speed of the ANC filters. The results showed that the NLMS approach performed better than the other three algorithms in terms of efficiency and quickness of convergence.

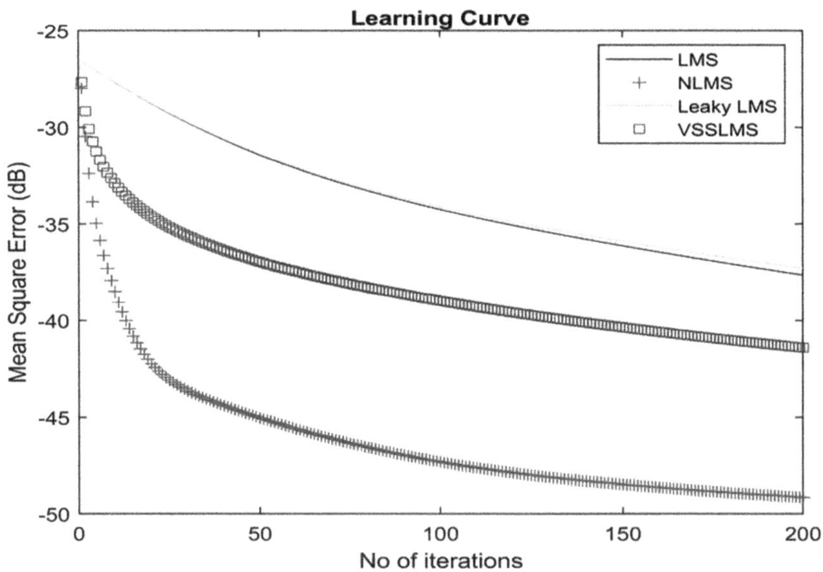

FIGURE 3.8 Contrast the convergence of four distinct algorithms: LMS, NLMS, LLMS, and VSS-LMS.

KEYWORDS

- **active noise control**
- **adaptive filters**
- **adaptive algorithms**
- **Internet of Things**
- **mean square error**

REFERENCES

1. Kuo, S. M., & Morgan, D. R. Active noise control: A tutorial review. *Proceedings of the IEEE* 1999, *87*(6), 943–973.
2. Sayed, A. H. *Adaptive Filters*. John Wiley & Sons, **2011**.
3. Kuo, S. M. Adaptive active noise control systems: Algorithms and digital signal processing (DSP) implementations. In *Proceeding of Digital Signal Processing Technology: A Critical Review* **1995**, *10279*, pp. 26–52.
4. Gwadabe, T. R., Salman, M. S., & Abuhilal, H. A comparison study between recursive inverse and iterative wiener filter algorithms. In: *Proceedings of EEECEGC*, **2013**, pp. 101–106.
5. Potdar, M. R., Meshram, M., Dewangan, N., & Kumar, R. Implementation of adaptive algorithm for PCG signal denoising. *IJIREEICE* **2015**, *3*(4), 33–42.
6. Haykin, S. S. *Adaptive Filter Theory*. Pearson Education India, **2002**.
7. Ren, A., Du, Z., Li, J., Hu, F., Yang, X., & Abbas, H. Adaptive interference cancellation of ECG signals. *Sensors* **2017**, *17*(5), 942.
8. Poularikas, A. D., & Ramadan, Z. M. *Adaptive Filtering Primer with MATLAB*. CRC Press, **2017**, p. 19.
9. Avalos, J. G., Sanchez, J. C., & Velazquez, J. Applications of adaptive filtering. *Adaptive Filtering Applications*, **2011**, *1*, 3–20.
10. Corinthios, M. Signals, systems, transforms, and digital signal processing with MATLAB. CRC Press, **2018**, p. 3.

CHAPTER 4

Introduction to Path Planning in an Unmanned Aerial Vehicles (UAV) Network

ANSHU VASHISTH* and BALRAJ SINGH

School of Computer Science Engineering, Lovely Professional University, Phagwara, Punjab, India

*Corresponding author

ABSTRACT

Drones, also known as unmanned aerial vehicles (UAVs), have transformed a number of industries including logistics, agriculture, disaster relief, and surveillance. In order to maximize operational effectiveness and mission success, this chapter explores the crucial area of path planning in UAV networks. Finding the most efficient and secure path for UAVs while taking into account limitations such as energy efficiency, collision avoidance, dynamic environments, and mission-specific goals is known as path planning. Emphasizing their adaptability to dynamic and diverse terrains, the chapter examines a number of algorithms and techniques essential to UAV path planning, such as graph-based, sampling-based, and metaheuristic approaches. Real-time decision-making and inter-UAV coordination are further improved by cutting-edge technologies like IoT integration and AI-driven optimization, allowing for smooth operations in networked environments. Challenges including communication barriers, scalability in multi-UAV networks, and environment adaptation are also covered in the text. It is explored how cutting-edge technologies like swarm intelligence, edge computing, and machine learning can support creative and flexible path-planning techniques. The chapter demonstrates how

IoT Potential for Green Energy Solutions. Abhishek Kumar, Ajay Roy, & Shamimul Qamar (Eds.)

DOI: 10.1201/9781779643599-4

UAVs may revolutionize efficiency, safety, and scalability through case studies in precision agriculture, urban surveillance, and disaster response. This chapter highlights the changing role of UAVs in redefining sectors, driven by developments in AI, collaborative algorithms, and intelligent systems, and concludes with insights into future trends.

4.1 INTRODUCTION

Unmanned aerial vehicles (UAVs), commonly known as drones, have grown in popularity and utility in a variety of applications, including surveillance, agriculture, disaster management, and communication. Path planning for UAVs in a networked environment is a significant factor that influences the efficiency and efficacy of UAV operations. This chapter presents an overview of UAV network path planning, concentrating on essential principles, problems, and cutting-edge technologies. Because of technological breakthroughs and a wide range of applications, the use of UAVs has grown quickly. Multiple UAVs collaborate to achieve specified objectives in UAV networks, demanding effective path-planning algorithms to enhance their performance. Path planning entails deciding on a series of actions that will allow a UAV to reach its destination while avoiding obstacles and sticking to stated limits. A UAV network is made up of many UAVs that communicate and work together to complete tasks. The network structure and communication protocols are critical in properly coordinating UAVs. Path planning is the process of determining an ideal or near-optimal path for a UAV to take from its starting point to a destination while taking terrain, obstructions, mission objectives, and energy restrictions into account. The path planning process seeks to optimize numerous goals, including lowering trip time, maximizing coverage, reducing energy usage, and avoiding collisions. UAVs operate in a dynamic environment with continually changing conditions like weather, obstructions, and other moving objects. The complexity of path planning and coordination develops exponentially as the number of UAVs in a network grows, necessitating scalable methods and methodologies.

4.1.1 THE ESSENCE OF PATH PLANNING

The essence of path planning in UAVs is determining the optimal or near-optimal path for a UAV to navigate from its initial position to a target destination while taking terrain, obstacles, mission objectives, and energy constraints into account. The goal of this procedure is to maximize the efficiency, effectiveness, and safety of UAV operations by carefully arranging a series of activities that allow the UAV to attain its target while avoiding obstacles and conforming to set limits. Path planning is critical for optimizing trip time, maximizing coverage, lowering energy consumption, and avoiding collisions, all of which are critical for effective UAV operations in a variety of applications and dynamic situations. Furthermore, because UAV networks comprise several UAVs working together to achieve specified goals, effective path-planning algorithms are required to coordinate their actions and optimize their aggregate performance. UAVs' meteoric development has altered a wide range of industries, including agriculture, surveillance, disaster management, logistics, and others [1, 2]. The growing use of UAV networks has increased the demand for sophisticated path-planning strategies. Path planning comprises determining the best route for a UAV to take from its starting point to a certain target while taking into account variables such as terrain, obstructions, mission objectives, and energy limits. According to an interactive analysis research, the usage of UAVs will increase by 2022 in terms of different applications for target missions indicated by a literature review [1]. According to Figure 4.1, the valuation of the UAV industry exceeded $15 billion by the conclusion of 2022.

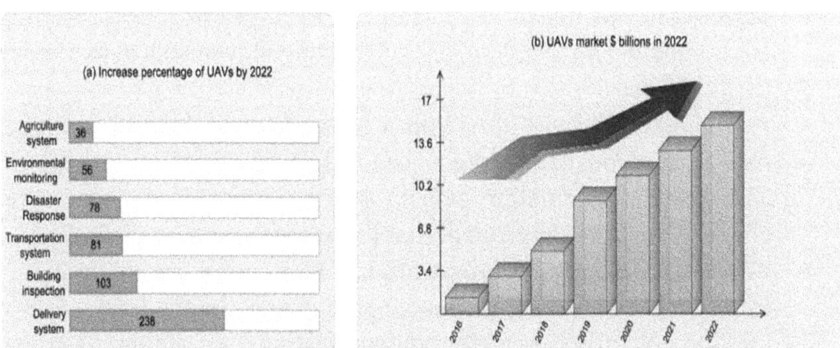

FIGURE 4.1 (a) Age (%) increase in the usage of UAVs by 2020 in various fields. (b) Market of UAVs in $ billions by 2022.

Source: Reprinted from Ref. [51]

4.1.2 OBJECTIVES

In this chapter, we aim to:

- provide an in-depth grasp of path planning in UAV networks in this chapter.
- emphasize essential topics and terms.
- investigate the specific issues of UAV path planning.
- discuss cutting-edge path-planning techniques and algorithms.
- provide insights into UAV path planning research's future trends and directions.

4.1.3 UAVS IN NETWORKED ENVIRONMENTS

In networked environments, UAVs refer to several UAVs that are interconnected and capable of communicating and collaborating to achieve specified goals. In this environment, UAVs do not operate in isolation, but rather as part of a wider network in which they can communicate information [3, 4], coordinate activities, and work together to complete tasks or missions. The following are key points about UAVs in networked environments:

- *Interconnectivity:* In a networked environment, UAVs communicate with one another via communication protocols, allowing them to share data and coordinate their motions or actions.
- *Collaboration:* The UAVs collaborate to achieve shared aims or objectives. This collaboration can include task distribution, sharing environmental information, or coordinating movements to maximize performance.
- *Information Sharing:* UAVs in a network communicate pertinent data such as position, status, sensor data, and mission objectives. This shared information aids in making educated decisions and charting the best course of action [5, 6].
- *Coordination and Synchronization:* UAVs coordinate their actions to minimize conflicts, guarantee smooth operations, and achieve mission objectives. Activity synchronization is critical for effective collaboration and optimal resource utilization.
- *Efficiency and Optimization:* The networked environment enables UAV operations such as path planning, resource allocation, and task distribution to be optimized to improve efficiency and meet overall mission objectives more efficiently.

- *Scalability:* The networked UAV environment can accommodate a rising number of UAVs, with each contributing to the overall effort while maintaining excellent communication and collaboration. Search and rescue missions, surveillance of large areas, monitoring of critical infrastructure, precision agriculture, environmental monitoring, and disaster response are examples of UAV applications in networked environments where multiple UAVs collaborate to achieve objectives efficiently and cover a larger area [7, 8].

4.1.4 IoT-ENABLED UAVS: AN ARCHITECTURAL OVERVIEW

UAVs have a layered structure that incorporates data collection, processing, and actuation. The collecting layer includes certain physical components that serve as onboard sensors for sensing or capturing data, such as temperature detection, light detection, pressure detection, and so on. These UAV hardware components gather data, which is subsequently processed by a control system, allowing UAVs to map, localize, and make decisions (see Figure 4.2). High-level functional components are depicted in Figures 4.3 and 4.4. Infrared devices provide a solution when there is a barrier or traffic in the path of UAVs by producing a warning and alarm buzzer. UAVs are equipped with cameras and video recorders to offer a visual representation of their environment. An electronic speed controller (ESC), a GPS module, sensors, a flight controller, and a battery are the essential components of an unmanned aerial vehicle (UAV). All of the components are linked and work together to increase production and precision.

- *ESC*: This circuit is critical in UAVs because it regulates the speed of electronic motors by converting DC power to AC power.
- *Global Positioning System Module*: This module aids UAVs in determining the position and time of data. UAVs may also give additional information such as latitude, longitude, and even compass information. This data is crucial for comprehending flying system navigation.
- *Flight Controller:* This module accepts GPS module input and operates on it.
- *Battery:* UAVs require electricity to function over lengthy periods.
- *Sensors:* Sensors in UAVs are broadly categorized into two groups. The first category comprises 3D photographs as well as LIDAR-captured thermal images. Thermal vision cameras are most commonly employed in agriculture and the environment to

detect illness and breed plants. When the surrounding environment is three-dimensional, all of these senses are required to make an appropriate choice, and they are also used [6].

- *Sensors and Data Collection:* UAVs are outfitted with a plethora of sensors, including cameras, LiDAR, GPS, and environmental sensors. During flight, these sensors capture real-time data, recording vital environmental and operational characteristics.

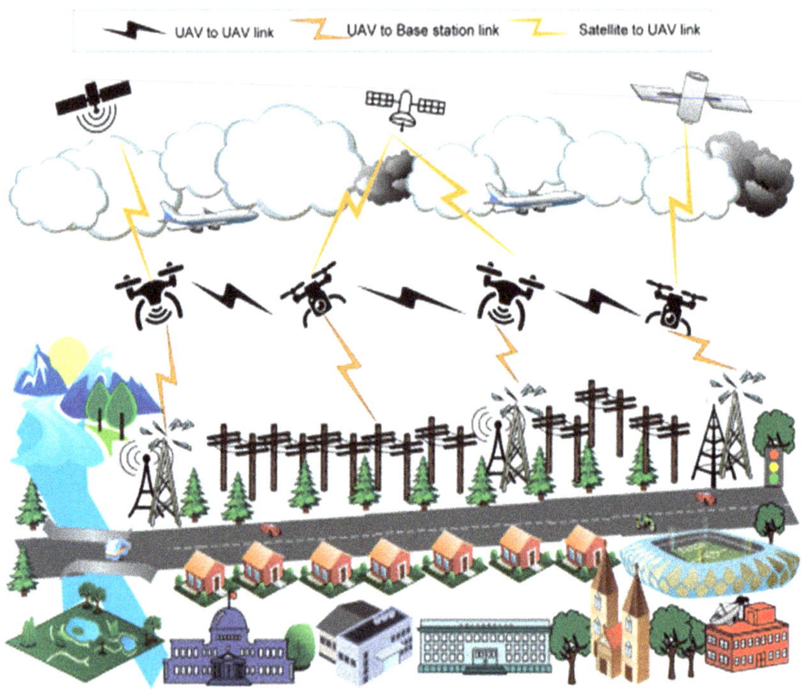

FIGURE 4.2 Architecture of UAVs.

Source: Reprinted from Ref. [51]

- *Onboard Data Processing:* Data is collected and processed onboard by embedded computer devices. This step entails data analysis, feature extraction, and the development of preliminary insights, as well as preparing the data for transmission and further study.
- *Communication Systems:* Advanced communication technologies on UAVs allow for flawless data sharing between the UAV and base stations. Satellite communication, radio frequency (RF), and cellular networks can all be used to do this.

- *Data Transmission:* Using communication networks, processed data is transferred in real-time to selected ground stations or cloud platforms. This allows for remote access to live data while the UAV is in flight.
- *Cloud Computing and Data Storage:* Data from the UAV is received and saved on cloud-based systems. Cloud computing allows for scalable data storage, processing, and sophisticated analytics while also serving as a centralized center for data administration and analysis.
- *Remote Monitoring and Control:* Through a user interface connected to the cloud platform, operators may remotely monitor UAV operations, modify flight settings, and update mission goals, assuring adaptive control and safety.

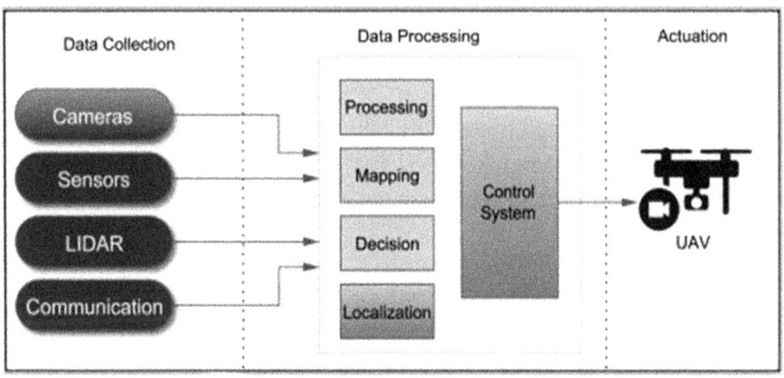

FIGURE 4.3 UAV components.

Source: Reprinted from Ref. [51].

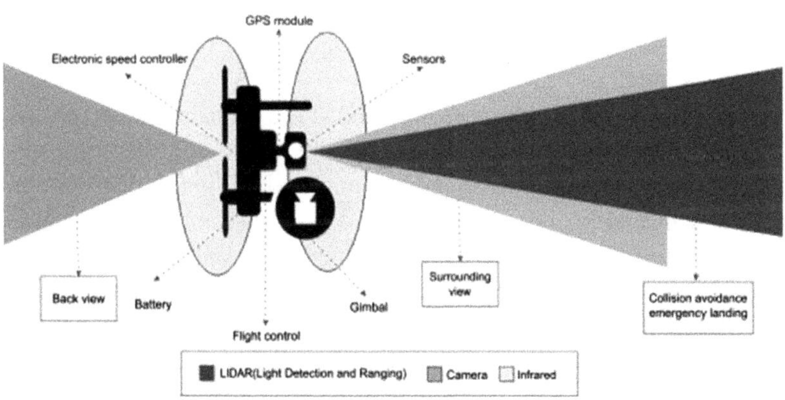

FIGURE 4.4 UAVs' components.

Source: Reprinted from Ref. [51].

4.2 FOUNDATION OF PATH PLANNING ALGORITHM

4.2.1 GRAPH-BASED ALGORITHMS

In the field of path planning for UAVs, graph-based algorithms are widely used to find optimal or near-optimal paths through a given environment. The environment and possible paths are represented as graphs by these algorithms, with nodes representing places or states and edges representing possible transitions or links between these locations. Here are some graph-based techniques that are commonly employed in UAV path planning [9, 10].

4.2.1.1 DIJKSTRA ALGORITHM

Dijkstra's algorithm calculates the shortest path between nodes in a weighted network, making it suited for UAV path planning where costs vary depending on parameters such as distance, obstructions, or energy consumption.

- *A (A Star) Algorithm*:* A* is a well-known heuristic search algorithm for finding the shortest path between nodes in a graph. It guides the search toward the goal node by using both the cost to reach a node (*g*) and a heuristic estimate of the cost to reach the goal (*h*).
- *Rapidly Exploring Random Trees (RRT):* RRT is a randomized method used in high-dimensional areas for efficient exploration and path planning. It builds a tree structure by randomly sampling the configuration space and growing into previously undiscovered locations.
- *Probabilistic Roadmap (PRM):* PRM is a probabilistic graph-based technique that samples random points in the configuration space, connects them to nearby points, and builds a graph that represents possible paths. It is very beneficial for high-dimensional planning.
- *Visibility Graphs:* These graphs generate a graph connecting visible points in the environment using line-of-sight visibility. Finding paths in this visibility graph can simplify UAV path planning, frequently resulting in efficient and collision-free routes.
- *D Algorithm*:* D* (Dynamic A Star) is an incremental search algorithm that can swiftly adjust the path in response to environmental

changes. It is appropriate for dynamic contexts where pathways must be changed as new information becomes available.

These techniques are critical in enabling UAVs to plan economical and safe trajectories by assessing the properties and restrictions of the environment in a graph-based representation. To efficiently design pathways for UAVs, an appropriate graph-based technique can be chosen depending on the unique application and needs [11, 12].

4.2.2 SAMPLING-BASED ALGORITHM

Because they give quick and frequently probabilistically complete solutions for traveling across large and high-dimensional areas, sampling-based algorithms are useful in UAV path planning. These methods produce a discrete set of points in the configuration space *via* random sampling and then construct a graph or tree structure connecting these sampled points to plan suitable courses for UAVs. Here are some sampling-based algorithms that are commonly employed in UAV path planning [1, 3]:

- *Rapidly Exploring Random Trees (RRT):* RRT is a popular UAV path-planning algorithm. It starts with a configuration and generates a tree iteratively by extending toward randomly picked locations in the configuration space. RRT explores the space effectively and determines a path from the starting point to the destination.
- **RRT (RRT Star)**:* RRT* is an RRT algorithm extension that optimizes the path by restructuring the tree structure to find a more optimal path to the goal. Its goal is to reduce overall path costs and give a near-optimal solution.
- *RRT-Connect*:* RRT*-Connect is an algorithm that boosts RRT*'s efficiency and convergence. It takes a bidirectional strategy, developing two RRT* trees from the start and goal places at the same time and connecting them to identify the best path.
- *Probabilistic Roadmap (PRM):* A roadmap is generated by sampling random points in the configuration space and connecting them if a collision-free path exists. It generates a graph representation of the space, allowing for fast path planning through the use of graph search techniques.
- *Visibility-Based Probabilistic Roadmap (VPRM):* The VPRM incorporates elements of visibility graphs and PRM. It creates a roadmap

by sampling sites and connecting them if they are visible to each other, making it an effective method for UAV path planning.

These sampling-based algorithms provide flexible and efficient solutions for UAV path planning in various scenarios, including environments with obstacles, dynamic conditions, and complex geometries. They contribute to the optimization of UAV trajectories, ensuring safe and effective navigation for a wide range of applications [13, 14].

4.2.3 METAHEURISTIC ALGORITHM

Metaheuristic algorithms are powerful optimization techniques that are extensively applied in UAV path planning to find high-quality solutions for complex problems. These algorithms are designed to efficiently explore and search through the solution space, often providing near-optimal or good solutions within a reasonable amount of time. Here are some commonly used metaheuristic algorithms in UAV path planning:

- *Genetic Algorithms (GAs):* GAs use ideas like selection, crossover, and mutation to imitate the process of natural selection. They are utilized in UAV route planning to evolve and optimize paths using fitness functions that take into account objectives including path length, obstacle avoidance, and energy consumption [15].
- *Particle Swarm Optimization (PSO):* Inspired by the social behavior of birds or fish, PSO is a type of optimization algorithm. It consists of a swarm of particles traveling across the solution space, altering their placements depending on their personal experiences as well as the best experiences of the entire swarm. PSO is used to optimize UAV trajectories based on a variety of objectives.
- *Ant Colony Optimization (ACO):* ACO is based on ant foraging behavior. UAV route planning with ACO entails imitating ant behavior to identify the optimal path by depositing pheromones on path edges and choosing paths depending on pheromone levels [16].
- *Simulated Annealing (SA):* SA is based on the metallurgical annealing process. It begins with an initial solution and iteratively updates it, tolerating poor solutions to avoid becoming caught in local optima. SA is used in UAV path planning to effectively explore the search space.

- *Genetic Programming (GP):* GP uses genetic algorithm principles to evolve programs that represent viable solutions. In UAV route planning, it is used to evolve algorithms or functions that generate optimum pathways based on predefined criteria [17, 18].

4.2.4 TECHNIQUES FOR UAV PATH OPTIMIZATION

See Table 4.1.

TABLE 4.1 Classification of AI Techniques

AI Technique	Description
Genetic algorithms	Evolves and optimizes UAV flight paths by simulating natural selection.
Reinforcement learning	Trains UAVs through trial and error, optimizing paths based on rewards.
Q-learning	Utilizes a Q-table to guide UAVs in making optimal decisions for path selection.
Neural networks	Analyzes historical data and mission requirements to predict optimal flight paths.
Ant colony optimization	Simulates the behavior of ants to optimize paths, considering pheromone levels.
Particle swarm optimization	Optimizes paths by simulating the social behavior of particles in a swarm.
Simulated annealing	A heuristic optimization technique for finding optimal paths by mimicking annealing in solids.
Tabu search	An optimization method that iteratively refines UAV paths by avoiding visited solutions.

4.3 KEY METRIC IN PATH PLANNING

Path planning for UAVs is a key process that entails establishing the best or near-best trajectory from a starting point to a goal while taking into account numerous objectives. The performance of UAV missions is frequently judged by how effectively the path planning fulfills specified key metrics and objectives, such as trip time optimization, energy efficiency, collision avoidance, and coverage maximization [19, 20].

4.3.1 TRAVEL TIME OPTIMIZATION

Travel time optimization is a key goal in UAV path design. It focuses on reducing the time it takes a UAV to reach its destination. Efficient path planning enables UAVs to navigate the environment quickly, which is especially important for missions requiring real-time response or time-sensitive activities. The path planner seeks to identify the shortest way to the destination while guaranteeing safety and adherence to mission requirements by taking into account elements such as path length, obstacle avoidance, and the UAV's speed capabilities [21].

4.3.2 ENERGY EFFICIENCY

Energy efficiency is a critical statistic in UAV operations, as it has a direct impact on the UAV's endurance and overall mission capability. Because UAVs often have limited onboard power, optimizing the approach to save energy consumption is crucial. Efficient path planning can aid in the reduction of energy-intensive movements, the consideration of wind conditions, and the selection of courses that need the least amount of energy. UAVs can increase their flight time, cover more regions, and complete more jobs in a single trip by conserving energy.

4.3.3 COLLISION AVOIDANCE

To protect the safety and integrity of the UAV and its surroundings, collision avoidance is a nonnegotiable goal in UAV path design. The path planner must guide UAVs so that they do not collide with barriers, other UAVs, or unexpected items in the environment. The route planner develops collision-free trajectories using sensors, obstacle detection algorithms, and real-time monitoring, allowing the UAV to navigate confidently while reducing the possibility of accidents or damage to the UAV and its surroundings [22].

4.3.4 MAXIMIZING COVERAGE

Maximizing coverage is an important goal in UAV operations, especially in applications such as surveillance, search and rescue, environmental monitoring, and agricultural inspections. The route planner's goal is to create a trajectory that permits the UAV to cover as much ground as feasible throughout its mission. This includes optimizing the flight path to scan the specified region effectively, adaptively modifying the course based on sensed data, and ensuring that all important regions are appropriately seen, photographed, or monitored to meet mission objectives.

To summarize, good UAV path design requires a careful balance of these essential measurements and objectives. For optimal mission performance, the ideal route should prioritize minimizing flight duration, enhancing energy efficiency, preventing collisions, and maximizing area coverage as needed. By tailoring the path planning process to prioritize and balance these objectives, UAV operations are more likely to be successful, efficient, and safe, resulting in the achievement of mission goals and objectives [23–25].

4.4 MAPPING THE TERRAIN: PATH REPRESENTATION TECHNIQUES

Mapping the terrain is a fundamental aspect of UAV operations, and path representation techniques play a crucial role in efficiently navigating and capturing the characteristics of the terrain. These techniques involve representing the environment in a way that allows the UAV to plan optimal paths, avoid obstacles, and achieve mission objectives. Here are several path representation techniques commonly used in UAV terrain mapping.

4.4.1 GRID-BASED REPRESENTATION

A grid-based representation separates the terrain into cells, each of which represents a distinct area. The UAV travels through these cells, which are labeled with information such as obstacles, terrain type, and other pertinent qualities. This model makes path planning easier by discretizing the environment into manageable units, allowing the UAV to design routes by traveling from one cell to the next [26].

4.4.2 POLYGONAL REPRESENTATION

The terrain is divided into polygons based on natural or manmade elements in polygonal representation. These polygons can represent a variety of properties such as land cover types (forests, water bodies, etc.) or specific areas of interest. UAVs can navigate between these polygons to plan pathways, allowing for efficient mapping and analysis of various terrain aspects.

4.4.3 GRAPH-BASED REPRESENTATION

A graph-based representation of the landscape represents key points (e.g., waypoints and barriers) and edges reflect possible connections or paths between these points. The UAV plots its course by exploring the graph, and picking the best paths based on criteria such as distance, elevation changes, and obstacle avoidance [27].

4.4.4 FEATURE-BASED REPRESENTATION

This method finds distinguishing features in the terrain, such as edges, corners, or critical spots. These features enable UAVs to plan courses that follow interesting or crucial points in the terrain, improving mapping accuracy and completeness.

4.4.5 WAYPOINT-BASED REPRESENTATION

A fundamental approach in UAV path planning and navigation is waypoint-based representation. Defining a set of waypoints or critical points within the environment that the UAV should visit or pass through throughout its mission is the goal of this strategy. These waypoints direct the UAV's trajectory and aid in the achievement of certain mission objectives. Here's a more in-depth look into waypoint-based representation in UAV path planning [28]:

- *Waypoint Selection and Generation:* Waypoints are picked strategically depending on mission requirements, objectives, and the geography or environment. These waypoints can be predefined based on existing knowledge, operator input, or dynamically produced utilizing algorithms that take terrain features, impediments, and mission objectives into account.
- *Spatial Distribution:* Waypoints are scattered over the area of interest most efficiently. The requirement to gain maximum coverage, avoid

obstructions, monitor certain places of interest, or follow a specified flying pattern to acquire desired data or imagery typically influences spatial distribution.

- *Connectivity of Waypoints:* Waypoints are linked together to establish a network or path that the UAV will follow. To capture the required data, connectivity between waypoints is built depending on the planned flight path, taking into account parameters such as obstacle avoidance, shortest path, or unique flight patterns.
- *Dynamic Waypoint Adaptation:* Waypoints may need to be adapted in real time in dynamic situations or evolving missions. Adaptive algorithms can change the waypoint distribution or connection dynamically in response to changes in the environment, new mission objectives, or sensor data obtained during the UAV's flight [29].
- *Considerations for Altitude:* Waypoints include information about the ideal altitude or elevation at which the UAV should approach or pass through them. This altitude data is crucial for optimizing the flight path of the UAV, especially in terrain with varied altitudes.
- *Real-time Monitoring and Adjustment:* The UAV's onboard electronics continuously monitor the UAV's position and status about the waypoints during the flight. Real-time modifications to the UAV's trajectory can be made to ensure that it stays on course and completes its mission successfully [30].

4.4.6 SPLINE CURVES AND INTERPOLATION

- *Spline Curves:* Spline curves are smooth, piecewise-defined curves that can be used to depict complex pathways in a variety of ways. They are commonly employed in UAV trajectory planning to ensure continuous motion of the UAV, which is required for smooth and effective flying. Reference [31] is an example of a spline curve. Bezier curves are defined by a series of control points that determine the curvature of the curve. They are useful for designing smooth pathways, and the UAV can follow a succession of Beziers curves to accomplish the required trajectory.
- *B-Spline Curves:* B-Spline curves are defined by a set of control points and basic functions. They are widely used for UAV path planning due to their ability to generate smooth and flexible curves, making them suitable for representing flight paths [32, 33].

- *Hermite Interpolation:* Hermite interpolation uses both the position and tangent information at control points to create a smooth curve between them. This is valuable for UAVs to smoothly navigate through waypoints with controlled orientation [34].
- *Cubic Interpolation:* Cubic interpolation, such as cubic splines, fits a cubic polynomial curve between data points, resulting in a smooth and continuous trajectory. It provides a balance between smoothness and complexity, making it suitable for UAV path planning [35, 36].

4.5 CHALLENGES IN UAV PATH PLANNING

Path planning for UAVs is a key process that entails charting a path across the environment to achieve a given goal while taking into account various restrictions and objectives. This procedure, however, is not without difficulties, particularly given the dynamic nature of settings, the requirement for good communication and coordination, and the scale constraints associated with networked UAVs.

4.5.1 ADAPTING TO DYNAMIC ENVIRONMENTS

Adapting to dynamic and unpredictable environments poses a significant challenge in the path planning of UAVs. Natural habitats are subject to rapid changes caused by various factors, including fluctuating weather conditions, moving objects like vehicles and pedestrians, and shifting terrain. To avoid collisions, navigate around obstructions, and optimize their routes, UAVs must dynamically adjust their planned courses in real time. Dynamic path replanning can be aided by using real-time data from multiple sensors, weather forecasts, and advanced machine learning (ML) algorithms. Algorithms that enable UAVs to make quick decisions based on changing conditions are critical for guaranteeing the safety and efficiency of UAV operations [37].

4.5.2 COMMUNICATION AND COORDINATION HURDLES

Effective communication and coordination among UAVs are critical to successful path planning, especially in scenarios involving numerous UAVs.

UAVs may transmit critical information regarding their trajectories, positions, and observations in real-time thanks to seamless connectivity. However, obstacles like as communication delays, limited bandwidth, and potential network congestion might stymie this crucial data transmission. Furthermore, it is difficult to ensure effective coordination to avoid conflicts, permit efficient work distribution, and avoid collisions. It is critical to design strong communication protocols and coordination algorithms that are resistant to communication uncertainty. Overcoming these obstacles is critical for successful multi-UAV path planning and coordination in a collaborative setting [38–40].

4.5.3 SCALABILITY ISSUES IN NETWORKED UAVS

As the usage of UAVs in networked environments increases, scalability emerges as a significant obstacle. The number of UAVs in the network increases the difficulty of path planning dramatically. As the number of UAVs increases, traditional path-planning algorithms may fail to accommodate the computing load and communication demands. As a result, planning timeframes are extended, resources are used inefficiently, and collision risks are increased. To address scalability difficulties, scalable algorithms and approaches that can handle a high number of UAVs must be developed. Hierarchical path planning methodologies, decentralized decision-making, and parallel processing are all viable answers to scalability issues and ensure that the path planning process stays effective and timely in networked UAV systems. Finally, tackling the issues of UAV path planning in dynamic situations, effective communication and coordination, and scalability in networked UAVs is critical for the successful integration and deployment of UAVs. To ensure safe, efficient, and scalable UAV operations in the present and future, research, innovation, and breakthroughs in these areas are critical [41].

4.6 EFFECTIVE COMMUNICATION FOR COLLABORATIVE PATH PLANNING

Collaborative path planning is an important part of optimizing the operations of UAVs in a variety of applications such as surveillance, delivery, and disaster response. Effective communication is critical for successful collaborative path planning for UAVs. In the context of UAV networks, this chapter investigates communication protocols and the distinction between decentralized and centralized communication [42].

4.6.1 COMMUNICATION PROTOCOLS IN UAV NETWORKS

In collaborative path planning scenarios, communication protocols are critical for facilitating data transmission and coordination among UAVs. These protocols specify the norms and practices for data transmission, allowing UAVs to communicate seamlessly. The following are some of the most prevalent communication protocols used in UAV networks:

4.6.1.1 MESSAGE QUEUING TELEMETRY TRANSPORT

It is a lightweight, open-source protocol intended for low-bandwidth, high-latency, or unstable networks. It enables effective communication by employing a publish-subscribe architecture in which UAVs can broadcast information about their state or intents while also subscribing to pertinent updates from other UAVs [43].

4.6.1.2 SIMPLE OBJECT ACCESS PROTOCOL

It is a technology that enables organized communication between UAVs through the use of XML-based messaging. It enables UAVs to transmit information about their course planning, status, or mission objectives in a standardized manner.

4.6.1.3 REPRESENTATIONAL STATE TRANSFER

It is an architectural style that employs a client-server communication mechanism that is stateless. UAVs may connect and share data using HTTP requests, making it a simple and commonly used protocol for UAV network communications [44].

4.6.1.4 COOPERATIVE AWARENESS MESSAGE PROTOCOL

It is a communication protocol commonly used in vehicular networks, adapted for UAVs. It allows UAVs to broadcast their position, speed, and other relevant information, enabling cooperative path planning and collision avoidance.

4.6.2 DECENTRALIZED VERSUS CENTRALIZED COMMUNICATION

In the context of collaborative path planning, communication can be organized in either a decentralized or centralized manner, each with its advantages and considerations.

4.6.2.1 DECENTRALIZED COMMUNICATION

Each UAV connects directly with other neighboring UAVs in a decentralized communication strategy. This communication style encourages adaptability, scalability, and toughness. Decentralized communication is especially beneficial when UAVs must react fast to changing situations or when a central coordinator is not viable or feasible. However, due to limited global knowledge, decentralized communication may result in higher communication costs, probable collisions, or inefficiencies [45, 46].

4.6.2.2 CENTRALIZED COMMUNICATION

In a centralized communication approach, a centralized coordinator manages and directs communication amongst UAVs. The central organization gathers data from all UAVs, processes it, and then distributes new plans or instructions to the UAVs. Centralized communication enables global awareness, effective coordination, and path optimization based on a comprehensive perspective of mission objectives. It does, however, introduce a single point of failure and reliance on the central coordinator, which can be a constraint in dynamic or complicated contexts.

4.6.3 UAV APPLICATIONS-BASED PROTOCOL

See Table 4.2.

TABLE 4.2 Classification of UAV Applications Based on Protocol and Their Limitations

Application Area	Specification	Protocols/Techniques Used	Limitations
Agricultural monitoring	Plants defects and growth monitoring, weeds, biochemical composition and pest detection.	It consists of a LiDAR sensor with multi and hyper spectral and camera, biological sensors.	The constraint poses significant challenges for imaginary spatial and temporal resolution and dependency on geometric and radiometric correction for dynamic weather.
Geological mapping	It includes processes such as tie point extraction, Ground Control Point measurement, and camera calibration orientation.	Sensor selection for different survey objectives, such as LiDAR for topography and vegetation cover, photogrammetry for 3D-models and ortho photos, and thermal camera as for subsurface features and heat loss.	Adverse weather conditions and potential technical issues with the UAV or sensors, i.e., camera triggering, etc.
Wildlife monitoring	Time variant wildlife monitoring technique based on GPS trackers and satellite sensors	Methods used are correlated random walk, biased random walk, and mower method. Apart from this presented multiagent exploration algorithm.	While satellite-based communication technology enables global coverage, there is a preference for centralized communication among UAVs.
Environmental monitoring	This research is intended at the environmental remote sensing for UAV optical frame imagining conjunction with air borne and satellite data.	Direct dereferencing using on-board raw GNSS data, Post Processed Kinematic single-frequency carrier-phase without *in situ* ground support. Spatial resolution was generated with photogrammetric software.	Limited precision due to atmospheric conditions, satellite geometry, and multipath faults, lack of ground control, inability to manage occlusions (caused by tall buildings, trees, or GNSS signal blockage) that caused dereferencing errors.
Infrastructure inspection	To provide efficient and accurate power line inspections, reducing the need for human inspections. Using AI and edge computing.	The deployed solution constitutes 5 phases–discovery, data collection, processing, a n d fault analysis. Apart from this, they use AI for object detection using R-CNN. It includes a 5G network to provide lower latency.	Fault-tolerant discovery is bound to only 71% which needs more improvement. Despite UAVs, human manpower is also required to attain and correct the fault line transmission.

TABLE 4.2 *(Continued)*

Application Area	Specification	Protocols/Techniques Used	Limitations
Disaster response	The tasks involved are search and rescue, situational awareness, surveillance, and monitoring of critical infrastructure, disaster response, human and wildlife conservation.	It uses sensing equipment such as radiation detection, ultraviolet and infrared spectrometer, ground penetration radar, sonar sensor, and hyper-spectral camera.	Less hover time due to battery power consumption requires the flight route optimization policy for such an investigating area. Apart from these, it also requires that it should maintain the swam nature to provide efficacy and livelihood in disastrous issues.
Military applications	The task involves assigning asset of multiple networks of UAVs for surveillance and monitoring of a geographical region for security purposes.	Equipment used are a Brushless DCmotor, GSM Module, GPS Module, and Raspberry Pi. Uses diamond-shaped network connection for covering large areas, apart from this it is also equipped with beam-steering antenna using a Butler matrix.	Because of the swarm network construction, it is prone to cyberattacks, collects impure data in bad weather, and has a restricted payload capacity. However, swimming has the benefit of being very adaptive to fluctuating circumstances.

4.7 RECENT ADVANCES AND FUTURE DIRECTIONS

Because of the combination of ML, adaptive algorithms, and artificial intelligence (AI), UAVs have made major advances in path planning. These innovations are transforming UAV path planning, allowing for more efficient, adaptive, and intelligent autonomous operations. We will examine previous breakthroughs, present trends, and promising future improvements in UAV path planning [47] in this part.

4.7.1 MACHINE LEARNING IN UAV PATH PLANNING

By allowing UAVs to learn and adapt depending on data and experiences, ML has emerged as a valuable method for optimizing UAV path planning. Recent developments include:

4.7.1.1 REINFORCEMENT LEARNING

UAVs can learn optimal pathways by interacting with their surroundings, earning incentives for good maneuvers, and altering their methods accordingly, thanks to RL algorithms. UAVs can adjust to changing conditions and unexpected obstacles thanks to this dynamic learning technique.

4.7.1.2 DEEP LEARNING FOR OBSTACLE DETECTION

Deep learning models, namely convolutional neural networks, are being used to detect and identify obstacles in the path of the UAV. This real-time obstacle recognition is critical for route planning to avoid collisions.

4.7.1.3 GENERATIVE ADVERSARIAL NETWORKS FOR SIMULATION

These are used to produce realistic simulations of environments, aiding the training and validation of UAV path-planning algorithms. These simulated settings aid in the optimization of path-planning tactics before the deployment of UAVs in the actual world.

4.7.2 ADAPTIVE ALGORITHMS AND AI INTEGRATION

Adaptive algorithms and AI integration are critical in improving UAV path planning by allowing real-time modifications and intelligent decision-making.

4.7.2.1 ADAPTIVE TRAJECTORY PLANNING

Algorithms that dynamically adjust UAV trajectories based on real-time data, such as weather conditions, traffic [48], or other UAVs, ensure optimized paths and enhance safety and efficiency.

4.7.2.2 MULTIAGENT COORDINATION WITH AI

Integrating AI for coordination among multiple UAVs allows for complex collaborative path planning. AI helps UAVs communicate, coordinate actions, and optimize paths collectively, ensuring efficient coverage of areas and avoiding collisions.

4.7.3 PROMISING FUTURE DEVELOPMENTS

The future of UAV route planning promises exciting possibilities, pushing the boundaries of technology and paving the way for expanded capabilities and applications [49, 50].

4.7.3.1 SWARM INTELLIGENCE FOR UAV NETWORKS

Using swarm intelligence concepts in UAV path planning, inspired by natural swarm behavior, could result in highly efficient and adaptive UAV networks capable of self-organization and robust performance.

4.7.3.2 EDGE COMPUTING FOR REAL-TIME DECISION-MAKING

UAVs may make real-time choices locally, lowering latency, and enabling rapid response to changing environmental variables during course planning by leveraging edge computing and edge AI.

4.7.3.3 HUMAN–UAV COLLABORATION

Integrating human intelligence into UAV path planning *via* user interfaces that allow operators or specialists to provide high-level advice and intervention, combining human insights with machine efficiency [51, 52], could be a future advancement.

4.8 CASE STUDIES AND PRACTICAL APPLICATIONS

UAVs have found diverse applications across several domains, showcasing the potential of efficient path-planning algorithms. In this section, we will delve into case studies and practical applications of UAV path planning in critical areas such as search and rescue missions, precision agriculture, and urban surveillance. UAVs equipped with cameras and thermal sensors assisted search and rescue teams by providing real-time aerial imagery, enabling more targeted and efficient rescue efforts [53–55].

4.8.1 PRECISION AGRICULTURE AND CROP MONITORING

UAVs have revolutionized precision agriculture by providing farmers with valuable insights into their crops' health, enabling precise resource allocation, and improving overall yield.

Case Study: Vineyard Monitoring in California, USA

UAVs equipped with high-resolution cameras and multispectral sensors are used to monitor vineyards in California's wine-producing regions. The UAVs take precise photographs of the plants, which aid farmers in detecting stress, illness, or nutrient deficits. This data is then analyzed by sophisticated algorithms, allowing farmers to make data-driven decisions about irrigation, fertilization, and pest control, resulting in improved crop health and yields.

4.8.2 URBAN SURVEILLANCE AND SECURITY

UAVs are increasingly being used for urban surveillance and security purposes, assisting law enforcement and improving public safety.

Case Study: Police Operations in London, UK

UAVs have been used by law enforcement authorities in London to monitor crowded events, regulate traffic, and prevent crime. UAVs with high-resolution cameras and live-streaming capabilities can monitor public gatherings and potential security risks in real-time. The capacity to modify flight paths dynamically allows for rapid deployment and observation of specified locations, allowing a proactive and effective reaction to security problems.

4.9 FUTURE PROSPECTS AND CONCLUSION

The practical applications of UAVs in path planning continue to grow, driven by technological developments and the growing demand for efficient, data-driven decision-making. As UAVs become more sophisticated and affordable, their potential to alter industries and save lives becomes clearer. The incorporation of AI, ML, and adaptive algorithms in UAV path planning is poised to substantially boost their capabilities, opening up new applications and addressing complicated challenges across multiple domains. These case studies demonstrate UAVs' versatility and effectiveness in path planning, offering a glimpse into the bright future of UAV technology.

KEYWORDS

- **unmanned aerial vehicles**
- **drones**
- **path planning**
- **surveillance**
- **efficiency**
- **obstacle avoidance**
- **dynamic environment**
- **safety**
- **collision avoidance**

- **agriculture**
- **coverage maximization**
- **effectiveness**
- **energy constraints**
- **trip time optimization**
- **UAV operations**

REFERENCES

1. Chen, H., Wang, X. M., & Li, Y. A survey of autonomous control for UAVs. *International Conference on Artificial Intelligence and Computational Intelligence, AICI* **2009**, *2*, 267–271.
2. Karim, S., Heinz, C., & Dunn, S. Agent-based mission management for UAVs. In: *Sensor Networks and Information Processing Conference*, Melbourne, Australia, **2005**, pp. 481–486.
3. Hasircioglu, I., Topcuoglu, H. R., & Ermis, M. 3-D path planning for the navigation of unmanned aerial vehicles by using evolutionary algorithms. In: *Proceedings of the 10th Annual Conference on Genetic and Evolutionary Computation*, **2008**, pp. 1499–1506.
4. Jaakkola, A., Hyyppä, J., Kukko, A., Yu, X., Kaartinen, H., Lehtomäki, M., & Lin, Y. A low-cost multi-sensoral mobile mapping system and its feasibility for tree measurements. *ISPRS Journal of Photogrammetry and Remote Sensing* **2010**, *65*(6), 514–522.
5. Aggarwal, S., & Kumar, N. Path planning techniques for unmanned aerial vehicles: A review, solutions, and challenges. *Computer Communications* **2020**, *149*, 270–299.
6. Gupta, L., Jain, R., & Vaszkun, G. Survey of important issues in UAVs communication networks. *IEEE Communications Surveys and Tutorials* **2016**, *18*(2), 1123–1152.
7. Miller, B., Stepanyan, K., Miller, A., & Andreev, M. 3D path planning in a threat environment. In: *Proceedings of the IEEE Conference on Decision and Control*, **2011**, pp. 6864–6869.
8. Chaitanya, Krishna, V., Harsha, M. S., Kumar, L. M., Rahul, V., Reddy, P. G. K., & Kumar, A. Smart parking management system. *In IOP Conference Series: Earth and Environmental Science* **2021**, *796*(1), 012014.
9. Lin, C., He, D., Kumar, N., Choo, K. K. R., Vinel, A., & Huang, X. Security and privacy for the internet of drones: Challenges and solutions. *IEEE Communications Magazine* **2018**, *56*(1), 64–69.
10. Javaid, A. Y., Sun, W., Devabhaktuni, V. K., & Alam, M. Cybersecurity threat analysis and modeling of an unmanned aerial vehicle system. In: *IEEE International Conference on Technologies for Homeland Security*, **2012**, pp. 585–590.

11. Kharchenko, V., & Torianyk, V. Cybersecurity of the internet of drones: Vulnerabilities analysis and IMECA based assessment. In: *Proceedings of IEEE 9th International Conference on Dependable Systems, Services, and Technologies*, **2018**, pp. 364–369.
12. Mozaffari, M., Saad, W., Bennis, M., & Debbah, M. Mobile Unmanned Aerial Vehicles (UAVs) for energy-efficient internet of things communications. *IEEE Transactions on Wireless Communications* **2017**, *16*(11), 7574–7589.
13. Fotouhi, A., Qiang, H., Ding, M., Hassan, M., Giordano, L. G., Garcia-Rodriguez, A., & Yuan, J. Survey on UAVs cellular communications: Practical aspects, standardization advancements, regulation, and security challenges. *IEEE Communications Surveys and Tutorials* **2019**, *21*(4), 3417–3442.
14. Lozano-Pérez, T., & Wesley, M. A. An algorithm for planning collision-free paths among polyhedral obstacles. *Communications of the ACM* **1979**, *22*(10), 560–570.
15. Hwang, Y. K., & Ahuja, N. A potential field approach to path planning. *IEEE Transactions on Robotics and Automation* **1992**, *8*(1), 23–32.
16. Sud, A., Andersen, E., Curtis, S., Lin, M. C., & Manocha, D. Real-time path planning in dynamic virtual environments using multiagent navigation graphs. *IEEE Transactions on Visualization and Computer Graphics* **2008**, *14*(3), 526–538.
17. Hayat, S., Yanmaz, E., & Muzaffar, R. Survey on unmanned aerial vehicle networks for civil applications: A communications viewpoint. *IEEE Communications Surveys and Tutorials* **2016**, *18*(4), 2624–2661.
18. Mozaffari, M., Saad, W., Bennis, M., Nam, Y. H., & Debbah, M. A tutorial on UAVs for wireless networks: Applications, challenges, and open problems. *IEEE Communications Surveys and Tutorials* **2019**, *21*(3), 2334–2360.
19. Tuna, G., Nefzi, B., & Conte, G. Unmanned aerial vehicle-aided communications system for disaster recovery. *Journal of Network and Computer Applications* **2014**, *41*(1), 27–36.
20. Yang, K., & Sukkarieh, S. 3D smooth path planning for UAVs in cluttered natural. *Environments* **2008**, 22–26.
21. Hrabar, S. D path planning and stereo-based obstacle avoidance for rotorcraft UAVs. In: *IEEE/RSJ International Conference on Intelligent Robots and Systems,* **2008**, pp. 807–814.
22. Samaniego, F., Sanchis, J., Garcia-Nieto, S., & Simarro, R. UAVs motion planning and obstacle avoidance based on adaptive 3D cell decomposition: Continuous space vs discrete space. *IEEE 2nd Ecuador Technical Chapters Meeting*, **2017**, pp. 1–6.
23. Lin, Y., & Saripalli, S. Sampling-based path planning for UAVs collision avoidance. *IEEE Transactions on Intelligent Transportation Systems*, **2017**, *18*(11), 3179–3192.
24. Nurimbetov, B., Adiyatov, O., Yeleu, S., & Varol, H. A. Motion planning for hybrid UAVs in dense urban environments. In: *IEEE/ASME International Conference on Advanced Intelligent Mechatronics*, **2017**, pp. 1627–1632.
25. Kwak, J., & Sung, Y. Autonomous UAVs flight control for GPS-based navigation. *IEEE Access* **2018**, *6*(c), 37947–37955.
26. Zhang, C., Liu, H., & Tang, Y. Analysis for UAVs heuristic tracking path planning based on target matching. In: *Proceedings of 9th International Conference on Mechanical and Aerospace Engineering,* **2018**, pp. 34–39.

27. Mengying, Z., Hua, W., & Feng, C. Online path planning algorithms for unmanned air vehicles. In: *Proceedings of IEEE International Conference on Unmanned Systems*, **2018**, pp. 116–119. ICUS.
28. Huang, D., Zhao, D., & Zhao, L. A new method of the shortest path planning for unmanned aerial vehicles. In: *Proceedings of IEEE 6th Data-Driven Control and Learning Systems Conference*, **2017**, pp. 599–605.
29. Perazzo, P., Sorbelli, F. B., Conti, M., Dini, G., & Pinotti, C. M. Drone path planning for secure positioning and secure position verification. *IEEE Transactions on Mobile Computing* **2017**, *16*(9), 2478–2493.
30. Mostafa, S. A., Mustapha, A., Shamsudin, A. U., Ahmad, A., Ahmad, M. S., & Gunasekaran, S. S. A real-time autonomous flight navigation trajectory assessment for unmanned aerial vehicles. *International Symposium on Agents, Multi-Agent Systems and Robotics*, **2018**, pp. 1–6.
31. Paredes, J. A., Álvarez, F. J., Aguilera, T., & Villadangos, J. M. 3D indoor positioning of UAVs with spread spectrum ultrasound and time-of-flight cameras. *Sensors* **2018**, *18*(1), 1–15.
32. Wang, X., & Meng, X. UAVs online path planning based on improved genetic algorithm with optimized search region. In: *Proceedings of the IEEE International Conference on Unmanned Systems*, **2019**, pp. 1–6.
33. Delavarpour, N., Koparan, C., Nowatzki, J., Bajwa, S., & Sun, X. A technical study on UAV characteristics for precision agriculture applications and associated practical challenges. *Remote Sensing* **2021**, *13*(6), 1204.
34. Tziavou, O., Pytharouli, S., & Souter, J. Unmanned Aerial Vehicle (UAV) based mapping in engineering geological surveys: Considerations for optimum results. *Engineering Geology* **2018**, *232*, 12–21.
35. Kabir, R. H., & Lee, K. Wildlife monitoring using a multi-UAV system with optimal transport theory. *Applied Sciences* **2021**, *11*(9), 4070.
36. Padró, J. C., Muñoz, F. J., Planas, J., & Pons, X. Comparison of four UAV georeferencing methods for environmental monitoring purposes focusing on the combined use with airborne and satellite remote sensing platforms. *International Journal of Applied Earth Observation and Geoformation* **2019**, *75*, 130–140.
37. Hildmann, H., & Kovacs, E. Review: Using unmanned aerial vehicles (uavs) as mobile sensing platforms (msps) for disaster response, civil security, and public safety. *Drones* **2019**, *3*(3), 1–26.
38. Benarbia, T., & Kyamakya, K. A literature review of drone-based package delivery logistics systems and their implementation feasibility. *Sustainability* **2022**, *14*(1), 360.
39. Utsav, A., Abhishek, A., Suraj, P., & Badhai, R. K. An IoT based UAV network for military applications. In: *International Conference on Wireless Communications, Signal Processing and Networking, WiSPNE,* **2021**, pp. 122–125.
40. Xu, C., Liao, X., Tan, J., Ye, H., & Lu, H. Recent research progress of unmanned aerial vehicle regulation policies and technologies in Urban low altitude. *IEEE Access* **2020**, *8*, 74175–74194.
41. Lekidis, A., Anastasiadis, A. G., & Vokas, G. A. Electricity infrastructure inspection using AI and edge platform-based UAVs. *Energy Reports* **2022,** *8*, 394–411.

42. Huang, H., Savkin, A. V., & Huang, C. Decentralized autonomous navigation of a UAV network for road traffic monitoring. *IEEE Transactions on Aerospace and Electronic Systems* **2021**, *57*(4), 2558–2564,
43. Hart, P. E., Nilsson, N. J., & Raphael, B. A formal basis for the heuristic determination of minimum cost paths. *IEEE Transactions on Systems Science and Cybernetics* **1968**, *4*(2), 100–107.
44. Karaman, S., & Frazzoli, E. Sampling-based algorithms for optimal motion planning. *The International Journal of Robotics Research* **2011**, *30*(7), 846–894.
45. Anderson, S., & Beard, R. W. Decentralized, cooperative path planning of unmanned air vehicles. *AIAA Journal of Guidance, Control, and Dynamics*, **2018**, *31*(2), 94–108.
46. Kirkpatrick, S., Gelatt, C. D., & Vecchi, M. P. Optimization by simulated annealing. *Science* **1983**, *220*(4598), 671–680.
47. Schertzer, E., & Julian, B. A* for fixed-wing UAV path planning in obstacle-rich environments. In: *IEEE/RSJ International Conference on Intelligent Robots and Systems (IROS)*, **2017**, pp. 1173–1180.
48. Steinberg, A. *Path Planning Algorithms for Unmanned Aircraft Systems with a Space-time Graph*. PhD dissertation., Florida Atlantic University, **2021**, pp. 53–76.
49. Liu, H., & Scherer, S. RRT*-Smart: An optimal sampling-based motion planning algorithm for robust decision-making. *Journal of Field Robotics*, **2017**, *34*(6), 1063–1085.
50. Vashisth, Batth, R. S., & Ward, R. Existing path planning techniques in Unmanned Aerial Vehicles (UAVs): A systematic review. In: *International Conference on Computational Intelligence and Knowledge Economy (ICCIKE),* Dubai, United Arab Emirates, **2021**, pp. 366–372
51. Vashisth & Batth, R. S. An overview, survey, and challenges in UAVs communication network. In: *International Conference on Intelligent Engineering and Management (ICIEM)*, London, UK, **2020**, pp. 342–347
52. Vashisth, A., Singh, B., & Batth, R. S. QMRNB: Design of an efficient Q-learning model to improve routing efficiency of UAV networks via bioinspired optimizations. *International Journal of Computer Networks and Applications (IJCNA)*, **2023**, *10*(2), 256–264.
53. Kaur, G., & Batth, R. S. Edge computing: Classification, applications, and challenges. In: *2nd International Conference on Intelligent Engineering and Management (ICIEM)*, **2019**, London, United Kingdom, 2021, pp. 254–259.
54. Kaur, G., Vashisth, A., & Batth, R. S. X-Ortho: Fuzzy rule-based expert system for diagnosing infective diseases of hinge joint knee. In: *International Conference on Automation, Computational and Technology Management (ICACTM)*, London, UK, **2019**, pp. 518–524.
55. Vashisth, A., Singh, B., Garg, R. et al. BPACAR: Design of a hybrid bioinspired model for dynamic collision-aware routing with continuous pattern analysis in UAV networks. *Microsystem Technologies* **2023**, *30*(4), 411–421.

CHAPTER 5

Impact of Internet of Things on Intelligent Transportation Systems

PANKAJ KUMAR KESHRI* and ABHISHEK KUMAR

School of Electronics and Electrical Engineering, Lovely Professional University, Phagwara, Punjab, India

*Corresponding author

ABSTRACT

Effective planning and time management are more important than ever in a time of fast-changing lifestyles, busy schedules, and an increasing need for instant access to information. Over 75 billion devices are expected to be online by 2025, highlighting the global trend toward pervasive connection and real-time communication. The Internet of Things (IoT), which is driving innovation in several industries, including transportation, is at the heart of this shift. Smarter, more sustainable and more efficient solutions are being made possible by the IoT's incorporation into local transportation networks, which are completely changing mobility. Applications of IoT in transportation are improving public transportation and efficiency, advancing smart city development, and encouraging intelligent mobility that tackles present and upcoming issues. IoT technology has the potential to completely transform the transportation industry as it develops and stays relevant, providing substantial advantages in terms of operational effectiveness, safety, and connectivity. This study examines how IoT is driving innovation in the transportation industry, emphasizing how it enhances public transit networks and advances the objectives of smart city projects.

IoT Potential for Green Energy Solutions. Abhishek Kumar, Ajay Roy, & Shamimul Qamar (Eds.)

DOI: 10.1201/9781779643599-5

5.1 INTRODUCTION

People are moving more rapidly due to our changing habits, hectic schedules, and an increasingly fast-paced way of life. We must therefore better organize our lives and make better use of our time. More than 75 billion gadgets are expected to be online by 2025, according to projections. There is no stopping the trend of people wanting to communicate with each other and have access to immediate information. The "Internet of Things" (IoT) significantly accelerates innovation in the local transportation sector and offers practical solutions. The transportation sector is increasingly utilizing the IoT, and there are several benefits associated with doing so to facilitate intelligent mobility [1–3]. IoT technology, which has a lower likelihood of becoming out of date soon, has enabled a revolution in the transportation sector. Enhancing public transportation is one of the primary objectives of the establishment of smart cities.

5.2 SMART TRANSPORTATION

Intelligent transportation systems employ a range of technologies to monitor, evaluate, and manage transportation networks in order to increase effectiveness and safety. The concepts of management, efficiency, and safety can be summarized as the three main components of smart transportation. We will put aside ideas of science fiction-style transportation for the time being. Stated differently, smart transportation makes use of state-of-the-art technology to enhance the convenience, safety, and affordability (for both the city and the individual) of traveling through urban areas. What new technology offers up these new possibilities? The growth of IoT devices and the advancement of 5G are the primary factors [4, 5]. By integrating inexpensive sensors and controls, the former enables remote control and management of nearly any physical machine. The latter provides the quick connectivity needed for real-time management and control of transportation networks with the least amount of latency.

Smart transportation is not just a concept for the future; systems are being developed in new locations based on the successes and failures of current implementations in various cities. Some of the cities that are implementing cutting-edge transportation technologies may surprise you at first. Naturally, smart transportation has been embraced by big cities with high levels of urban intelligence, such as New York City.

In addition to being deployed on buses and trains, IoT devices can also be integrated into the city's infrastructure as shown in Figure 5.1. The public transportation system can be continuously visible to transport regulators thanks to sensors on highways, lamps, platforms, bus stops, railroad lines, and other sections of travel routes [6–8]. Transportation authorities may be able to provide improved services to their citizens without sacrificing operational effectiveness if this is paired with the data collected from commuters' gadgets.

FIGURE 5.1 Smart transportation system.

Source: Reprinted with permission from Ref. [14]. Copyright © 2023, The Author(s), under exclusive licence to Springer-Verlag GmbH Germany, part of Springer Nature

5.3 BENEFITS OF SMART TRANSPORTATION SYSTEM

Integrating IoT technology with logistics and transportation systems yields several benefits. These benefits have been discussed in the following sections.

5.3.1 CONNECTED ALL MODES OF TRANSPORTATION

As seen in Figure 5.2, it is essential to all modes of transportation, including land, air, and sea travel. The cloud servers or other servers that transmit data to networks connect the smart devices (sensors, processors) that go into building each component of these transportation fields. Compared to simply moving from one place to another, IoT transportation is safer, more environmentally friendly, and more convenient. For example, a smart car can do several things at once, such as communication, entertainment, navigation, and more dependable, effective transportation. Numerous wireless technologies, such as Bluetooth, Wi-Fi, 4G, 5G, intelligent traffic systems, and even other cars, are used to connect the vehicle to the Internet.

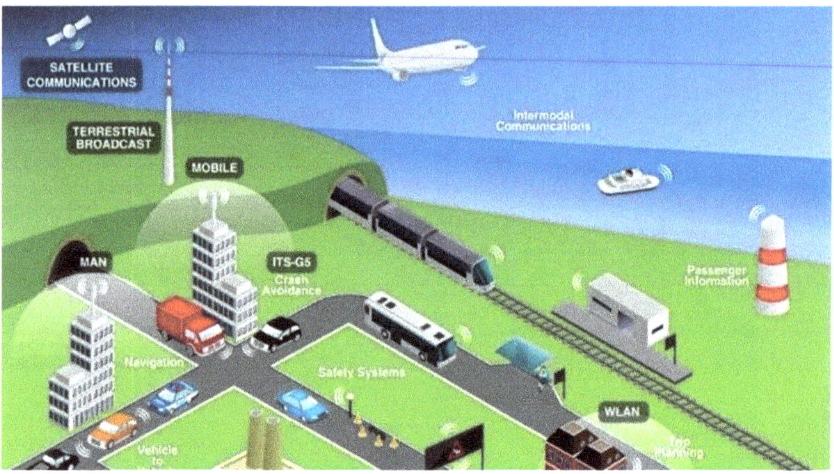

FIGURE 5.2 IoT-based various mode of transportation system.

Source: Adapted from Ref. [15].

5.3.2 TRAFFIC MONITORING AND COLLISION AVOIDANCE

A vehicle's exterior or interior can have sensors installed to prevent lane departure and to continuously monitor objects on all sides, as seen in Figure 5.3. Networks for public transportation and logistics are centrally integrated through a traffic control system based on the number of vehicles [9, 10]. To optimize travel, automated real-time decision-making is made possible by the IoT component of transportation, which extends outside the vehicle to communicate with others. For example, a traffic monitoring camera notifies the nearest traffic control facility of an accident or traffic conjunction and provides current traffic conjunction information to other vehicles in the vicinity, advising them to take an alternate route. Under dangerous and lethal conditions, users can reroute the optimal paths. The largest area of the transportation industry where IoT technology adoption is thought to be most prominent is traffic management. CCTV cameras are generating an immense amount of data, in the order of millions of billions of gigabytes, pertaining to traffic and vehicles. The traffic management facilities receive this information in order to monitor the vehicles more closely and impose fines on the drivers of vehicles that violate the law. Traffic and patrolling officers are aided in efficiently managing traffic and reducing the likelihood of accidents by a number of IoT applications, including smart parking, autonomous traffic signal systems, and smart accident aid.

FIGURE 5.3 Traffic monitoring system.

Source: Reprinted from Ref. [13]. Open Access

With the use of GPS trackers, transportation companies can quickly obtain a vehicle's location as well as its specifics and statistics. This allows transportation companies to monitor their most important assets in real time. IoT devices have the ability to monitor a user's whereabouts as well as their activities while driving and can give details about their driving patterns and idle time. IoT has helped fleet management systems lower their operating, gasoline, and maintenance costs. Real-time tracking, in terms of transportation monitoring, has made it much simpler to put wise decisions into practice by allowing drivers to spot problems with their vehicles right away and take preventative measures as needed.

5.3.3 COST EFFICIENT

Smart transportation can reduce expenses because it makes better use of the resources at hand, resulting in cheaper energy costs, less preventative maintenance, and fewer resources dedicated to accidents. When affordable public transportation is competitive with the ownership of a private automobile, users can also save money. Increasing the vehicle's driving range while minimizing fuel consumption leads to improved daily operational profits.

5.3.4 ENVIRONMENTAL CONSIDERATIONS

Transportation and the environment have always been linked. Whether it is the combustion engines of today or the steam cars of the past that ran on coal and wood, transportation has a detrimental effect on the planet's resources and atmosphere. The implementation of smart transportation technology enables communities to utilize their existing resources more responsibly. Despite ongoing scientific developments in alternative energy sources for transportation, mass transit, which is more environmentally friendly than private vehicles, is often impractical in certain regions, diminishing its widespread adoption in the US and other countries. But with the efficiency gains that come from smart transport solutions, modern public transportation might become profitable for more groups of people. As urban transportation technology advances in major cities, the tried-and-true techniques can be adopted and disseminated to areas that desire the advantages of a smart city.

5.3.5 SUPPLY CHAIN RESILIENCY

The world's supply lines are susceptible to interruption, as demonstrated by global catastrophes like the Coronavirus epidemic. Utilizing autonomous transportation for commodities has the potential to potentially save lives in scenarios where workers and drivers are unwell, and relocating poses a threat to public health. The development of an autonomous supply chain that can deliver essential products like food and emergency supplies without putting human drivers at danger may depend on initiatives like Wyoming's linked vehicle project, which is powered by smart, city-to-city transportation, and logistics networks, as shown in Figure 5.4.

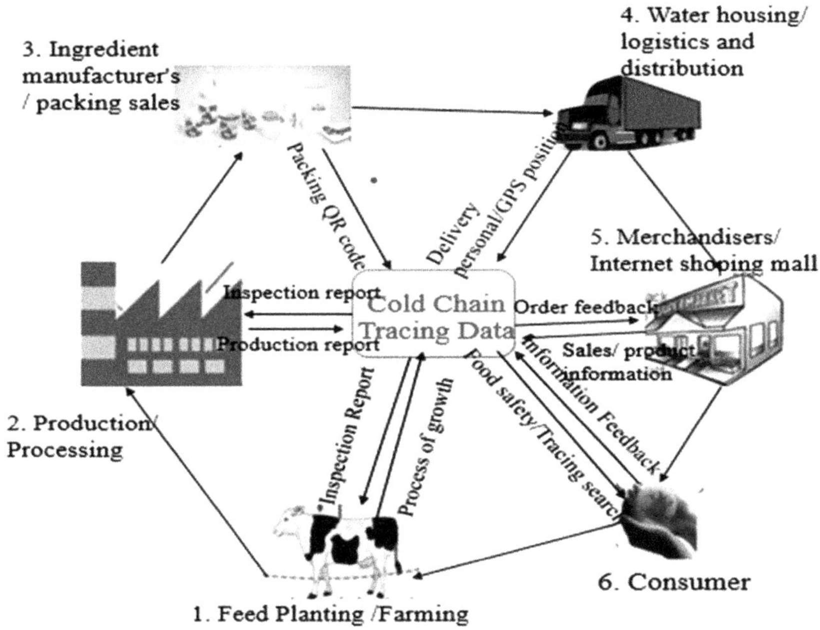

FIGURE 5.4 Smart supply chain system [10].

Wherever safety and effectiveness can be increased through automation, artificial intelligence, and robotics, the human supply chain labor can be minimally supplemented. The good news is that professionals anticipate that these developments will encourage the formation of more jobs, safer jobs specifically, as developers, technicians, analysts, and administrators work to promote and maintain the breakthroughs.

5.3.6 OPTIMIZING TRANSPORT ROUTES

By deploying IoT sensors at bus stops and train platforms, transportation authorities can precisely estimate the volume of foot traffic there. Based on passenger count, they can identify which station or stop will need the most pickups. With this information, different routes can be assigned with the appropriate number of vehicles. Additionally, they may determine the quickest routes by using the traffic data acquired by the sensors positioned on street lights, at traffic signals, and along roadways [11, 12]. Bus drivers and train conductors can receive real-time updates on the most efficient routes between two locations, allowing them to swiftly transport passengers to their destinations. Additionally, train schedules can be optimized based on passenger volume at each stop. By dispatching trains at a higher frequency than other stations, stations with high foot traffic can be efficiently supplied. In order to ensure maximum speed without compromising safety, sensors on resist tracks can be used to accurately coordinate the traffic on the railway network. As a result, regulators will be able to use both the infrastructure and their cars efficiently. In the long run, this system's knowledge can be utilized to plan new transportation routes and launch additional bus and train services. In addition to aiding in decisions about infrastructure, this information may be useful for city planning and management.

5.3.7 SECURITY

The susceptibility of smart cities to cyberattacks is a main concern among detractors. After all, as the globe becomes more linked, cyberattacks on key infrastructure, such as Internet-connected power grids and banking systems, have almost become standard practice among criminals and even nation-states. But the attacks themselves are not innovative; only the means are. Long before computers existed, banks, power grids, and other crucial infrastructure, such as transportation, were susceptible to physical attacks. When vehicles and infrastructure are integrated, networked, and autonomous, physical risks such as thieves stealing automobiles, terrorists using vehicles as weapons, and bad guys taking captive public transportation can all be reduced. Cyberattacks provide a greater danger than the physical threats mentioned above, but they are considerably easier to counter. Cyberattack risk can be reduced through the use of updated software, virtual private network tunnels for encrypted communications, and other multilayered security measures. This indicates that smart transportation for intelligent cities can improve the general safety of modern public transit by lowering the likelihood of both physical and cyberattacks.

In addition to enhancing security, IoT in transportation also supports public transportation administration by offering a variety of smart solutions. Innovative vehicle logistical solutions, passenger information systems, automatic fare collecting, and integrated ticketing are some of them. These remedies aid in traffic congestion management and public transportation. IoT has made real-time management of public transportation conceivable. This has made it easier for transportation authorities to communicate with passengers and deliver essential information *via* passenger information displays and mobile devices. IoT has clearly improved the safety and effectiveness of public transportation.

5.4 DISADVANTAGES OF SMART TRANSPORTATION

Although smart city transportation systems offer numerous benefits, there are also several drawbacks that may come to light during their implementation. These issues typically revolve around power usage and sensible data handling. Many sensors are needed for smart cities, and all of those sensors need power. Sensors affixed to moving objects need batteries for this. Static sensors almost always need to be connected to the city's electrical system, even though they may be able to run on solar power. Powering so many devices is a challenging issue because of the enormous number of sensors that are necessary for the globe to migrate to smart cities (estimated in the trillions).

Even with sensors connected to the electrical grid, the demand for raw materials such as copper surpasses the level of production that the global population is used to. Beyond questions of authority, the use of personal data on the Internet is a contentious issue in today's society. The energy that powers smart cities is data. The attitudes and behaviors of the populace will need to change, despite the fact that a large portion of the necessary data is anonymous when compared to online data. Cars will need to collect positional data, and citywide sensors will need to passively collect the signals that smartphones send out all day long. Enacting responsible laws and procedures for data management regardless of anonymity will be necessary for smart cities.

5.4.1 BUDGETING

For the purpose of gathering, moving, and transforming data, smart transportation systems often need a strong network infrastructure. The upfront costs for 5G IoT servers, long-range antennae, and sensors can be high. Fortunately, not every project necessitates having all the infrastructure in place right away. In regions with cellular coverage, IoT sensors can be gradually installed. As the system's value and funding increase, more areas can be covered over time.

5.4.2 MANAGING DATA

Large amounts of data can be produced by smart transportation networks. Without adequate organization, this data is lost or, worse yet, clogs the network as a whole. To guarantee the success of smart transportation deployment, it is essential to ensure that the appropriate technology infrastructure is in place. Verify that your network is equipped to collect the necessary data to meet your objectives. Assess whether your network has the capacity to support the 50 HD CCTV cameras needed for monitoring and redirecting traffic. Experts in cellular networks might be crucial in ensuring that your network can support the results you want.

5.4.3 MOBILITY

The majority of smart transportation use cases are by nature highly mobile. To secure the maximum return on investment and operational productivity for these systems as people, systems, and devices migrate, dependable connectivity and extensive wireless coverage become a critical necessity.

5.4.4 MAINTAINING PRIVACY AND SECURITY

Data gathering produces the benefits that smart transportation networks offer. If improper implementation is used, maintaining good data privacy might be difficult. Attackers may seriously harm people and property in the real world if there was weak cyber security. Through careful planning, wise financial management, and innovative technology alliances, many of these drawbacks can be avoided.

5.5 WHICH AREA OF PUBLIC TRANSPORT IS SEEKING IoT GROWTH

According to projections and observations, large investments are mostly seen in smart communities, transportation technology, and energy efficiency/power consumptions. Regarding the prioritization of investments, the majority of public transportation providers are digitizing crucial processes both inside their own internal vertical value chains and with their horizontal partners. Additionally, transportation businesses are adding cutting-edge data-based services and improving their product offering with digital functions.

5.6 CONCLUSION

The IoT enables new, high-tech services. IoT services improve the efficiency and reliability of transportation services by utilizing preexisting data in a flexible manner. Operators can communicate schedule changes to the public, while passengers are kept informed of the precise arrival time and location of their vehicle. Without a doubt, IoT services benefit both users and operators. Among the uses of IoT in the transportation industry are security, self-driving cars, traffic management, and real-time tracking. The IoT will keep improving transportation for as long as there is a future. Artificial intelligence, holograms, virtual assistance, and augmented reality are just a few of the cutting-edge technologies that will be included in the intelligent public transit system in the future. Additionally, 6G, which offers improved frequency, sustainability, and connection, is replacing 5G. To give commuters the best experience possible, such technology integration will be a game changer.

KEYWORDS

- **IoT**
- **sensors**
- **GPS**
- **6G**

REFERENCES

1. Jamjoom, L., & Alshmarani, A. A wireless controlled digital car lock for smart transportation. In: *15th IEEE Learning and Technology Conference. ISSN: 17805861*, **2018**, pp. 46–51.

2. Prabhu, B., Antony, A. J., & Kumar B. N. A research on smart transportation using sensors and embedded systems. *International Journal of Innovative Research in Computer Science & Technology (IJIRCST),* **2017**. 5.
3. Dweik, A. A., Muresan, R., Mayhew, M., & Lieberman, M. IoT-based multifunctional scalable real-time enhanced road side unit for intelligent transportation systems. In: *IEEE 30th Canadian Conference on Electrical and Computer Engineering (CCECE),* Windsor, **2017**, pp. 1–6.
4. Sutar S. H., Koul, R., & Suryavanshi, R. Integration of smart phone and IOT for development of smart public transportation system. In: *International Conference on Internet of Things and Applications (IOTA),* Pune, **2016**, pp. 73–78.
5. Herrera-Quintero, L. F., Banse, K., Vega-Alfonso, J., & Venegas-Sanchez, A. Smart ITS sensor for the transportation planning using the IoT and Bigdata approaches to produce ITS cloud services. In: *8th Euro American Conference on Telematics and Information Systems (EATIS),* Cartagena, **2016**, pp. 1–7.
6. Leal, A. G., Santiago, A., Miyake, M. Y., Noda, M. K., Pereira, M. J., & Avanço, L. Integrated environment for testing IoT and RFID technologies applied on intelligent transportation system. In: *IEEE Brasil RFID,* Sao Paulo, **2014**, pp. 22–24.
7. Yuan, X., Su S., & Chen, H. A graph-based vehicle proposal location and detection algorithm. *IEEE Transactions on Intelligent Transportation Systems* **2017**, *18*(12), 3282–3289.
8. Chong, H. F., & Ng, D. W. K. Development of IoT device for traffic management system. In: *IEEE Student Conference on Research and Development (SCOReD)*, **2016**.
9. Mukhtar, A., Xia, L., & Tang, T. B. Vehicle detection techniques for collision avoidance systems: A review. *IEEE Transactions on Intelligent Transportation Systems.* **2015**, *16*(5), 2318–2338.
10. Rakhonde, M. A., Khoje, S. A. K., & Komati, R. D. Vehicle collision detection and avoidance with pollution monitoring system using IoT. In: *IEEE Global Conference on Wireless Computing and Networking (GCWCN)*, **2018**, pp. 75–79.
11. Jain, N. K., Saini, R. K., & Mittal, P. A review on traffic monitoring system techniques. *Soft Computing: Theories and Applications*. Springer: Singapore, **2019**, pp. 569–577.
12. Li, J., Xiang, Y., Fang, J., & Wang, W. Research on multiple sensors vehicle detection with EMD-based denoising. *IEEE Internet of Things Journal*. **2019**, *6*, 6262–6270.
13. Pal, P. & Pali, L. Congestion free analysis for emergency vehicles response in tri-city (Panchkula-Chandigarh-Mohali) using LTE-A. Modelling, Measurement and Control A. 2018. 91. 66-72. 10.18280/mmc_a.910206.
14. Karthikeyan, H., Usha, G. A secured IoT-based intelligent transport system (IoT-ITS) framework based on cognitive science. Soft Comput 28. 2024, 13929–13939. https://doi.org/10.1007/s00500-023-08410-7
15. Abdur Razzaq, M., Shaikh, R., Qureshi, M., Baig, M. A. Internet of Things (IoT) Applications: An Overview. 2017

CHAPTER 6

Commercial Fleet Sustainability: IoT Tracking Progress With Environmental Monitoring

ASHISH A. DESAI*, GOVIND SINGH PATEL, S. B. HERWADE, G. V. PUJARI, and V. S. HAJARE

Department of Automation & Robotics, Sharad Institute of Technology, College of Engineering, Yadrav, Ichalkaranji, Kolhapur, Maharashtra, India

*Corresponding author

ABSTRACT

Sustainability in the commercial fleet is crucial for reducing environmental impact and improving operational efficiency. This study examines the potential benefits of Internet of Things (IoT) technology for sustainability through data-driven decision-making and real-time monitoring. The objectives include lowering carbon emissions, improving fuel efficiency, ensuring compliance with environmental regulations, and enhancing safety. IoT connects devices and systems to monitor fuel use, greenhouse gas emissions, and indoor air quality (IAQ). Advanced sensors lower risks and ensure compliance with environmental standards by giving real-time information on things like temperature management and issuing alerts for hazardous conditions like high CO_2 levels or driver fatigue. By integrating IoT, fleet operators may improve driver safety and lower accident rates while cutting expenses, optimizing resource use, and satisfying regulatory requirements. IoT-based IAQ systems, for example, can identify low air quality in driver cabins and initiate preventive measures, such as modifying ventilation or taking breaks, to lower the dangers associated with fatigue.

IoT Potential for Green Energy Solutions. Abhishek Kumar, Ajay Roy, & Shamimul Qamar (Eds.)

DOI: 10.1201/9781779643599-6

This chapter also looks at how IoT-enabled smart energy management systems can be implemented and evaluates how well they work to improve indoor air quality, lower energy use, and boost customer happiness. IoT technologies can help fleet operators tackle issues like air pollution and fuel inefficiencies while promoting environmental sustainability.

6.1 INTRODUCTION

The Internet of Things (IoT) refers to the network of interconnected devices, objects, and systems that communicate and exchange data with each other over the Internet. IoT enables seamless connectivity and data sharing, leading to automation, real-time monitoring, and control. IoT makes it easier for devices to connect and communicate with one another, facilitating effective data sharing and collaboration. IoT makes it possible to monitor and manage many factors in real-time, enabling proactive decision-making and quick reaction to changing circumstances. IoT improves safety and security by making it possible to identify potential threats or breaches to send out prompt alerts and take action. IoT enables personalized and connected experiences for customers, tailoring products and services based on their preferences and behaviors. The IoT enables effective management of infrastructure and systems and facilitates the creation of sustainable and intelligent solutions by reducing resource consumption. IoT has a big impact on healthcare since it makes it possible for remote patient monitoring, wearable health tracking devices, and effective healthcare delivery.

Commercial fleet accidents are unquestionably a severe problem that cannot be ignored. Poor air quality and driver fatigue are two of the most common causes of these accidents. According to studies, fatigue is more likely to have an impact on drivers who work in extremely polluted environments, which could limit their driving skills and increase the likelihood of collisions. Furthermore, the National Highway Traffic Safety Administration (NHTSA) reports that heavy trucks were involved in 4951 fatal crashes in 2018, an increase of 1% from the previous year. According to the same data, the bulk of these accidents happened during the day and on workdays. Therefore, to decrease the likelihood of an accident, it is essential to address the issue of air quality and driver safety. When particular limits are exceeded, IoT-based health and air quality monitoring systems can produce alerts or warnings. For instance, a warning can be given to the driver and fleet management if the cabin's carbon dioxide (CO_2) levels reach harmful levels, directing them to take preventative measures like opening windows, changing the air conditioning, or taking rest breaks. IoT-based approach enables the integration of sensors and communication systems to continuously monitor indoor air quality (IAQ) parameters such as temperature, humidity, CO_2 levels, volatile organic compounds (VOCs), and particulate matter (PM). By collecting real-time data from these sensors, the system can assess the air quality. Many authors might present the implementation details of the IoT-based IAQ and smart energy management system, discussing the hardware components, communication protocols, and data processing techniques employed. They may also evaluate the system's performance, considering factors such as energy savings, IAQ improvement, and user satisfaction. Overall, the chapter likely contributes to the existing body of knowledge on IoT applications in IAQ monitoring and smart energy.

6.2 LITERATURE REVIEW

The idea of utilizing IoT technology to improve air quality control is explored to make learning settings healthier and more supportive. The IoT-enabled system focuses on controlling the dynamic air quality in several classrooms of a school. Sensors are used to continually track variables affecting air quality, including temperature, humidity, CO_2 levels, and VOCs. These sensors are linked to a central control system, which performs real-time analysis of the data gathered. The system may dynamically modify ventilation and air purification systems to enhance air quality in each room based on the data analysis. This ensures that throughout the exposed to clean, healthy air. Additionally, the system can send pertinent stakeholders, like building managers or school administrators, real-time feedback, and notifications so that they can take preventative action if air quality deviates from permitted levels. The IoT-based strategy has several advantages, including better student health, higher academic results, and energy efficiency. The technology creates a healthier atmosphere, lowering the risk of respiratory disorders and other health problems, by continuously monitoring and improving air quality. Additionally, it helps create a more welcoming and conducive learning atmosphere, which benefits students' focus and well-being. Overall, the chapter introduces the idea of implementing dynamic air quality control and utilizing IoT technology. The objective is to provide smart and healthy learning environments that put well-being and academic achievement first using real-time data analysis and intelligent control technologies [1].

IAQ and energy management are crucial factors in commercial buildings to ensure a healthy and comfortable environment while optimizing energy consumption. Maintaining good IAQ is essential for the well-being and productivity of occupants, while efficient energy management reduces costs and environmental impact. Strategies for IAQ and energy management include real-time monitoring of IAQ parameters, such as temperature, humidity, CO_2 levels, VOCs, and PM, and implementing control measures to optimize ventilation and air purification systems. This helps create a healthier indoor environment and promotes energy efficiency. Continuous data collection, analysis, and control enable building operators to make informed decisions, address comfort issues, and reduce energy wastage, contributing to a sustainable and comfortable commercial building environment [2].

The various studies investigate the application of IoT technology to improve IAQ and optimize energy management in HVAC (heating, ventilation, and air conditioning) systems also mentioned, the use of IoT-enabled sensors to monitor IAQ factors such as temperature, humidity, CO_2 levels, VOCs, and particle matter (PM). These sensors collect real-time data on IAQ and provide insights into the quality of the air within a facility. Furthermore, the chapter emphasizes the use of IoT technology for smart energy management in HVAC systems. The technology may optimize HVAC operations to reduce energy usage while maintaining target comfort levels by evaluating the acquired data. These methods allow for the detection of patterns, trends, and anomalies in IAQ and energy usage, resulting in more informed decision-making and optimization tactics. The chapter also examines the possible advantages of integrating IoT-based IAQ and smart energy management systems. Improved IAQ, higher energy efficiency, lower operational costs, and increased occupant comfort and well-being are among the benefits. The suggested approach attempts to promote healthier indoor environments while optimizing energy usage in buildings by integrating IoT-enabled sensors and data analytics [3].

6.3 OBJECTIVE OF COMMERCIAL FLEET SUSTAINABILITY: IoT TRACKING PROGRESS WITH ENVIRONMENTAL MONITORING

1. **Reduce carbon emissions:** Monitor and reduce greenhouse gas emissions from commercial fleet.

2. **Increase fuel efficiency:** Reduce costs and environmental effects by improving fuel usage and operational efficiency.
3. **Make sure environmental compliance:** Comply with environmental laws and regulations to avoid penalties and contribute to protecting the environment.
4. **Enhance data-driven decision making-making:** Emphasize the importance of real-time data from sensors and IoT technology in making informed sustainability decisions.

6.3.1 METHODOLOGY

The implementation of IoT tracking devices in commercial fleets has had a substantial impact on fuel consumption and emissions management for logistics companies. These devices enable real-time monitoring and analysis of data related to engine performance, route efficiency, and driver behavior. By collecting such comprehensive data, logistics companies gain valuable insights into the overall fuel efficiency of their fleets. This not only allows for the optimization of fuel usage and cost reduction but also contributes significantly to the reduction of greenhouse gas emissions.

IoT tracking progresses with environmental monitoring and monitoring systems into commercial fleet. These systems detect and monitor the fleet's performance, energy usage, and emissions using sensors and GPS technology. This not only assists businesses in analyzing and optimizing fleet operations but it also allows them to track their progress in decreasing their environmental effect. Commercial fleet operators can make substantial gains toward a more sustainable and environmentally friendly transportation business by implementing IoT tracking and environmental monitoring technologies. In essence, IoT tracking devices play a crucial role in revolutionizing fuel consumption and emissions management in commercial fleets. The continuous monitoring and analysis provided by these devices empower logistics companies to make informed decisions, reduce environmental impact, and enhance overall operational efficiency [6, 7].

Figure 6.1 illustrates the IoTs monitoring and tracking devices for business fleet cars. These small devices can provide real-time information on a variety of issues, including fleet speed, fuel efficiency, and engine health. They can also track the fleet's whereabouts and provide alerts in the event of unauthorized use or theft. With this information, fleet operators may make informed decisions to increase overall operational efficiency and minimize fuel use, resulting in lower carbon emissions. Finally, the installation of IoT tracking and environmental monitoring systems can help to make the transport industry greener and more sustainable.

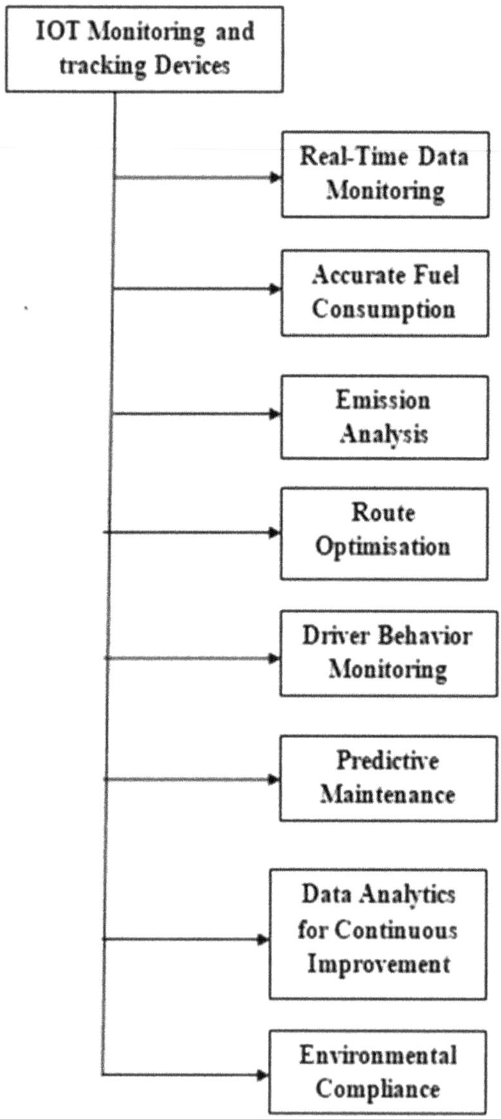

FIGURE 6.1 IoT monitoring devices.

The use of IoT tracking devices in logistics companies has modified real-time monitoring analysis of fuel usage and emissions data. This technology has various advantages that help business fleets improve their efficiency, sustainability, and cost-effectiveness:

6.3.1.1 REAL-TIME DATA MONITORING

Real-time data monitoring provides the company with up-to-date information on the fuel consumption and emissions of their fleet. This allows them to rapidly discover any anomalies or problems that are causing inefficiency. With real-time data, the organization can make quick modifications and take proactive measures to optimize fuel usage, resulting in financial savings and a lower environmental effect. Furthermore, the ability to monitor emissions in real time allows the company to maintain compliance with environmental requirements while also demonstrating to stakeholders and customers their commitment to sustainability.

6.3.1.2 ACCURATE FUEL CONSUMPTION TRACKING

Accurate fuel usage tracking is critical for businesses looking to improve efficiency and cut expenses. Real-time fuel usage data allows for the detection of abnormalities or faults that are generating inefficiencies, allowing for fast modifications and proactive efforts to optimize fuel consumption and generate significant cost savings. Furthermore, exact fuel usage tracking enables real-time monitoring of emissions, ensuring compliance with environmental requirements and demonstrating to stakeholders and customers the company's dedication to sustainability. IoT devices play an important part in this process by integrating fuel flow sensors and OBD-II sensors to properly monitor and track fuel consumption. This information is critical for logistics organizations in determining how effectively their cars utilize fuel, identifying patterns, and identifying areas for improvement.

6.3.1.3 EMISSIONS ANALYSIS

Emissions analysis with IoT devices enables businesses to precisely monitor and measure the quantities of pollutants generated by their cars in real time. Businesses may actively assess and analyze environmental effects by integrating exhaust gas sensors and emission control systems, allowing them to make educated decisions to reduce emissions. The IoT-driven strategy enables the rapid diagnosis and correction of issues or malfunctions that may contribute to excessive emissions, ensuring compliance with environmental requirements and reducing the fleet's overall carbon footprint. Furthermore, this technology enables logistics organizations to acquire extensive insights into the composition of exhaust fumes, including contaminants like carbon monoxide (CO) and nitrogen oxides (NOx). Real-time emissions analysis enables businesses to manage their environmental effect proactively, resolving pollution concerns and demonstrating a commitment to sustainability. IoT devices contribute to a full understanding of the fleet's environmental performance by utilizing exhaust gas analyzers and other emission-monitoring sensors. In addition to environmental benefits, the integration of IoT devices speeds the reporting process, providing regulatory agencies with accurate and up-to-date emissions data. This guarantees that emission requirements and laws are met, reaffirming corporations' commitment to environmental responsibility. Overall, including IoT in emissions research is an important step toward a greener, more sustainable future in the transportation sector.

6.3.1.4 ROUTE OPTIMIZATION

Route optimization, facilitated by IoT devices in the automotive sector, brings substantial benefits in efficiency, cost savings, and environmental sustainability. Real-time data from sensors and GPS tracking devices empowers fleets to navigate the most efficient routes, reducing fuel consumption, and saving time and money. This contributes significantly to a decreased carbon footprint, aligning with environmental sustainability goals. GPS sensors enable precise tracking of fleet routes, allowing logistics companies to strategically optimize delivery routes, minimizing both fuel usage and emissions, leading to reduced operational costs and a smaller carbon footprint. Beyond cost and emission reductions, IoT-enabled route optimization alleviates traffic congestion, enhances overall efficiency, and improves customer satisfaction, contributing to an improved company reputation. The integration of IoT in route optimization is a win–win solution for both businesses and the environment, promoting a more sustainable and eco-friendly transportation system. Additionally, businesses benefit from real-time data and analytics, enabling informed decisions, improved delivery times, increased productivity, and safer transportation practices. Ultimately, IoT in route optimization shapes a more efficient, eco-friendly, and safer future for the transportation industry.

6.3.1.5 DRIVER BEHAVIOR MONITORING

Monitoring driver behavior is crucial to ensuring workplace safety, utilizing sensors such as accelerometers and gyroscopes to identify critical behaviors such as abrupt braking or quick acceleration, both of which are indicators of unsafe driving practices. This real-time data are quickly communicated to operators or supervisors, allowing for quick intervention or remedial action. The incorporation of these sensor technologies enables businesses to protect their employees by proactively averting accidents and nurturing a safer and more efficient work environment. IoT devices are critical in enabling driver behavior monitoring. Various aspects of a driver's activities are tracked and analyzed using sensors and data-gathering devices, ranging from speed and acceleration to braking patterns and adherence to traffic rules. These data, which may be accessed in real time or retrospectively, provide insights into the driver's performance and behavior on the road. This complete monitoring strategy improves driving safety, efficiency, and overall driving practices,

6.3.1.6 PREDICTIVE MAINTENANCE

IoT devices contribute to predictive maintenance practices by monitoring numerous metrics linked to fleet health. Proactively identifying and addressing faults not only improves fleet reliability but also ensures optimal fuel efficiency. Furthermore, the real-time data collected by IoT devices enables maintenance personnel to more efficiently schedule repairs and replacements, minimizing downtime and overall costs. This comprehensive monitoring system also aids in the prevention of severe failures or accidents by recognizing possible problems before they worsen. Overall, IoT devices play an important role in enhancing fleet longevity and performance in the transportation industry.

6.3.1.7 DATA ANALYSIS FOR CONTINUOUS IMPROVEMENT

Regular examination of data collected by devices that track is critical to the success of sustainability programs. This practice enables businesses to identify trends, patterns, and areas for improvement, enabling data-driven decisions and focused plans to further reduce their carbon footprint. Furthermore, such research provides insights into the effectiveness of applied interventions, allowing for changes for optimum environmental impact. Furthermore, data analysis supports the development of relevant goals and the tracking of progress toward their achievement. Companies can measure their level of sustainability and discover areas for improvement by comparing key performance indicators to industry benchmarks. This information is critical for effectively communicating sustainability activities to stakeholders such as customers, investors, and regulators.

Furthermore, data analysis aids in the identification of cost-cutting options and the optimization of resource allocation, resulting in greater operational efficiency and profitability. Overall, including data analysis in sustainability programs is critical for businesses seeking to effect significant change, improve environmental performance, and fulfill their sustainability objectives.

6.3.2 SCHEMATIC REPRESENTATION OF FUEL AND AIR QUALITY MONITORING IN COMMERCIAL FLEETS

The integration of several components and sensors is depicted schematically in the fuel and air quality monitoring system of commercial fleets. First, a fuel level sensor that gauges the amount of fuel in the car's tank is part of the fuel monitoring system. The fuel level data are collected and sent to the cloud platform by the node MCU 8266, which is coupled to this sensor. The air quality monitoring system, on the other hand, is made up of sensors that find contaminants such as PM, nitrogen dioxide (NO_2), and CO. These sensors are positioned carefully within the exhaust system of the car to measure the emissions.

For instance, a fleet could have both an air quality and fuel monitoring system installed as part of a smart city operation. Fleet managers would be able to monitor and optimize charging schedules thanks to the fuel level sensor, which would measure the battery levels in each bus and relay the data to the cloud platform. In the meantime, the air quality sensors would continuously track each bus's emissions, giving real-time information on pollution levels and assisting in the identification of any fleets that could need repairs or modifications to lessen their negative effects on the environment [4, 5].

The intended objectives include building real data and analytics, displaying data and creating alarms for odd conditions, and optimizing Fleet usage and maintenance while enhancing driver safety and health. Here is a description of how these goals can be reached.

6.3.2.1 COLLECTING AND ANALYZING DATA

Gather information on driving safety factors such as accident histories, near-miss situations, and metrics for driver behavior (such as jarring braking, rapid acceleration, and speed). Utilize the sensors in moving fleets to assess air pollution, temperature, humidity, and other pertinent variables. Analyze the data gathered to find trends, patterns, and relationships between air quality levels and driver safety occurrences. To provide insightful conclusions, use statistical analysis, machine learning, or other analytics approaches.

6.3.2.2 ALERTS AND VISUALIZATION

Create a user-friendly interface with dashboards or other graphical representations that clearly illustrate the data and insights. When uncommon or dangerous conditions are found, such as high levels of air pollutants or instances of reckless driving behavior, use the interface to generate real-time alerts or notifications. Ensure that the alerts are clear, usable, and sent to the proper parties, such as drivers, fleet managers, or safety supervisors.

6.3.2.3 FLEET OPTIMIZATION AND SAFETY

Make use of the data analysis's findings to improve the way that fleets are used and maintained. It may be necessary to do this by identifying areas that need to be improved, such as by minimizing idle time, putting regular maintenance schedules in place, or suggesting driver training programs based on recognized harmful behaviors. Utilize information on fuel usage, driving habits, and air quality to develop tactics that encourage eco-driving and fuel efficiency, which can lower expenses and have a positive influence on the environment. To address driver well-being and reduce tiredness-related risks, incorporate driver health monitoring technologies, such as heart rate monitors or fatigue detection algorithms.

6.3.2.4 KNOWLEDGE AND INSTRUCTION

Create instructional resources, such as info-graphics or training modules, to inform fleet managers, drivers, and other interested parties of the significance of air quality for driver safety and wellbeing. Promote safe driving habits, highlight the importance of air quality, and promote behavioral changes that enhance driver safety and health through organizing awareness campaigns or training sessions. These goals can be met through coordinating data gathering, analysis, visualization, alerts, fleet optimization, and awareness programs to increase awareness, improve driver safety, optimize Fleet use, and promote overall health and well-being. Finally, IoT-based commercial fleets aspire to establish new business models and prospects, such as the development of digital services and products. Connected fleet can enable new types of logistics and supply chain management, providing greater visibility and control over the transportation of commodities

It is vital to keep in mind that the specific sensors, microcontroller, and connectivity modules utilized can change depending on the project's needs, the possibilities on the market, and the intended features. The connections between these parts and how they work together will be shown in Figure 6.2.

6.4 IoT PLATFORM AND ALGORITHM

Specialized features and data management are made possible by classifying an IoT platform into two modes for fuel and air quality monitoring, respectively.

- Integrate IoT Arduino Cloud with Node MCU 8266 for air monitoring
- Integrate Ubidots with Node MCU 8266 for Fuel Monitoring

Distinction between Ubidots employing Node MCU 8266 and Node MCU 8266-based IoT-based Arduino cloud. The main differences between an IoT-based Arduino cloud with Node MCU 8266 and Ubidots with Node MCU 8266 are in the associated platforms and functionality. An IoT-based Arduino cloud with Node MCU 8266 offers connectivity and data management options for IoT applications. It is a cloud platform designed specifically for Arduino projects. On the other hand, Ubidots is a comprehensive platform for IoT development that provides a wide range of features, including data visualization, analytics, and integration with various IoT devices, including Node MCU 8266. The specific requirements and objectives of the ongoing project would ultimately determine which option was best.

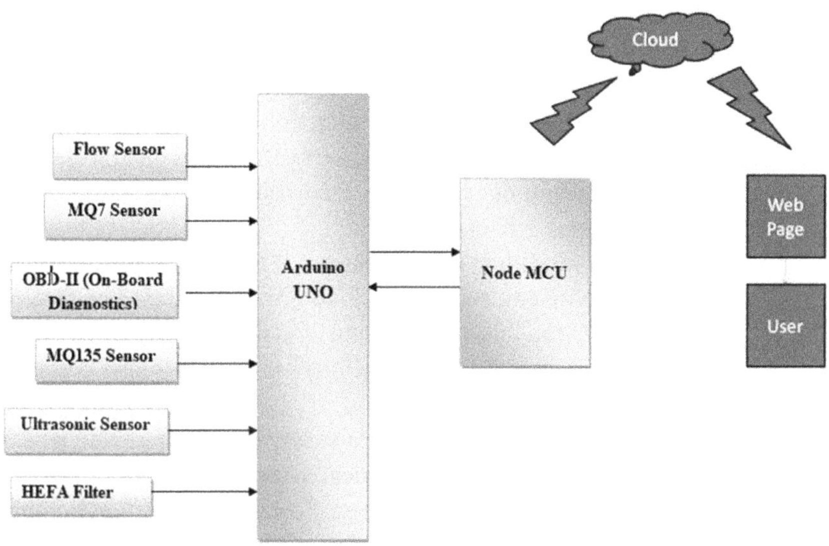

FIGURE 6.2 Schematic representation of air and fuel quality monitoring in commercial fleet.

6.4.1 INTEGRATE IoT ARDUINO CLOUD WITH NODE MCU 8266 FOR AIR MONITORING

Air quality monitoring in commercial fleets

1. To develop real data and analytics that can be used to raise awareness about the link between driver safety and air quality.
2. To visualize data and generate alerts for any unusual conditions using a user-friendly interface.

3. To optimize fleet usage and maintenance, reduce fuel consumption, and improve driver safety and health.

6.4.1.1 HARDWARE ELEMENTS FOR AIR QUALITY MONITORING

Here are some typical hardware elements used in IoT-based air monitoring systems for commercial fleet, while precise hardware specifications may vary depending on the implementation:

1. **Gas sensor:** Gas sensors are essential for identifying and monitoring airborne contaminants. Specific gases, such as CO, NO_2, PM, and VOCs, may be monitored using a variety of gas sensors.
2. **Microcontroller/Single board computer:** The system's central processing unit is frequently a microcontroller or single board computer, such as an Arduino, Raspberry Pi, or comparable device. It manages communication with other components and gathers data from the gas sensors.
3. **Connectivity modules:** IoT systems need connectivity modules to deliver data to distant servers or cloud infrastructures. The system may connect to the Internet and transfer data in real-time via these modules, which may be Wi-Fi, cellular, or other wireless communication modules.
4. **Power source:** The system needs a steady source of power, which might come from the fleet's electrical system or an external power source like a battery or power adapter. To ensure stable functioning, additional power management and conditioning components may be added.
5. **Data storage and transmission:** The system may make use of storage devices like SD cards or flash memory to store and send the gathered air-quality data. In addition, data can be sent to a central server or cloud platform via wireless data transmission modules or protocols like Wi-Fi or Bluetooth.
6. **Display and user interface:** A display module or user interface, such as an LCD screen or LED indicators, may be used in some systems to give the driver real-time feedback or alerts regarding air quality levels or system status.
7. **Enclosure and mounting:** To enhance longevity and dependability, an adequate enclosure and mounting mechanism may be used to

shield the hardware components from the elements and vibrations experienced during fleet operation.

The requirements and limitations of the IoT-based air monitoring system for trucks are crucial and may vary based on the hardware design and selection. Accuracy, power consumption, size, and cost are a few additional variables that may be taken into account while choosing components.

A comprehensive review on IAQ monitoring systems for enhanced public health, on the other hand, typically investigates the many technologies, methodologies, and applications associated with IAQ monitoring. It highlights the significance of IAQ in public health as well as the potential health hazards linked with poor IAQ.

6.4.1.2 CONCEPTUAL MODEL FOR AIR QUALITY MONITORING SYSTEM

Figure 6.3 indicates conceptual model for air quality monitoring system several essential elements and procedures for measuring, analyzing, and improving the quality of the air are often included in a conceptual model for air quality monitoring. Here is a summary of the key components of such a model at a high level.

1. **Sensors and data collection:** Setting up a network of sensors that can measure different contaminants and environmental factors is the initial stage in air quality monitoring. PM, NO_2, sulfur dioxide (SO_2), ozone (O_3), CO, VOCs, temperature, humidity, and other sensors are examples of what can be measured by these devices. The monitoring system is built on the data that these sensors acquire.
2. **Data storage and transmission:** The sensors' acquired data needs to be safely sent and kept in storage. This may entail sending real-time data to a centralized database or cloud storage via wireless communication technologies, such as IoT devices.
3. **Data processing and analysis:** To understand the air quality conditions, the data must be processed and analyzed after it has been gathered and saved. The data can be analyzed using a variety of algorithms and statistical techniques to spot trends and abnormalities.
4. **Calculating the Air Quality Index (AQI):** The observed data are frequently converted into an Air Quality Index (AQI) to provide

the general public with useful information. A numerical scale called the AQI is used to rank the severity of air quality, from good to hazardous. It aids in understanding how the current air quality affects health.

5. **Visualization and reporting:** User-friendly interfaces, such as those seen on websites or mobile applications, can be used to display the analyzed data and AQI results. Stakeholders and the general public can quickly understand trends and patterns in air quality thanks to data visualization.
6. **Alerting and public communication:** The system should have an alerting mechanism to notify the public, appropriate authorities, and stakeholders in the event of high pollution levels or hazardous situations. This might be done by SMS notifications, app notifications, or open announcements.
7. **Regulatory compliance:** Monitoring of air quality frequently plays a crucial part in ensuring regulatory compliance. The information may be used by government organizations to enforce pollution control laws, set emission limits, and create strategies to enhance air quality.
8. **Mitigation and control measures:** The conceptual model should cover the implementation of mitigation and control measures in addition to monitoring. The data analysis can be used to determine the best course of action for reducing pollution sources, enhancing ventilation, and improving overall air quality.
9. **Public awareness and education:** Part of a successful air quality monitoring program is teaching the general public about the significance of air quality, its effects on health, and personal actions that can help create cleaner air.

Air quality monitoring system in Figure 6.3 comprises the following:

1. **Gas sensors:** To identify the presence of dangerous gases like CO and smoke particles in the air, the air filtration system has gas sensors. These sensors continuously check the air quality, and they notify the user when it exceeds safe levels.
2. **Three indication lights:** There are three indication lights in the air filtration system: red, yellow, and green. Bad air quality is shown by the red light, continuing purification is indicated by the yellow light, and satisfactory air quality is indicated by the green light.

3. **High-efficiency particulate air (HEPA) filter:** The air filtration system is outfitted with a HEPA filter that collects and eliminates airborne tiny particles like smoke, pollen, and dust. Up to 99.97% of airborne particles as small as 0.3 microns can be removed by the HEPA filter.
4. **Air purification:** By continuously passing the air through the HEPA filter, the air filter system filters the air. The system automatically modifies its speed and intensity of filtration following the gas sensors' measurements of the air quality.
5. **User-friendly design:** The air filtration system is made to be easy to use and maintain. Using the simple control panel on the device, users may adjust the settings and monitor the air quality data.

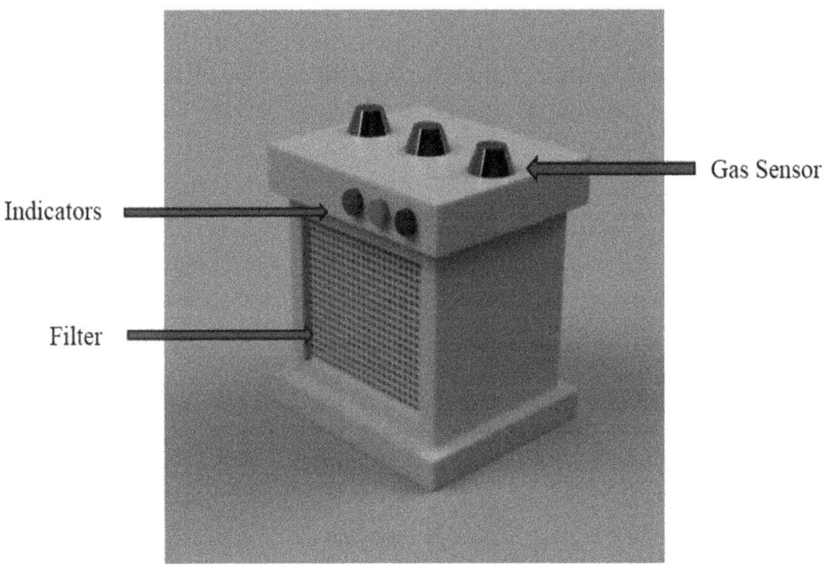

FIGURE 6.3 Air quality monitoring system.

By utilizing these technologies, real-time data on air quality and fuel levels can be collected and analyzed efficiently. This enables proactive measures to be taken, such as adjusting fuel consumption or addressing air pollution sources promptly. Overall, the combination of the IoT Arduino Cloud and Node MCU 8266 has proven to be an effective solution for monitoring and managing air and fuel conditions. Additionally, the Cloud and Node MCU 8266 allow for remote monitoring and control, eliminating the need for manual data collection and reducing the risk of human error. This technology also enables data to be accessed and analyzed from any location, providing a comprehensive view of air and fuel conditions across multiple sites or regions. Furthermore, the flexibility and scalability of the cloud and Node MCU 8266 make it possible to easily expand the monitoring system as needed, ensuring continuous and accurate data collection for optimal decision-making. Effect.

Figure 6.4 shows Node MCU 8266 for air monitoring The combination of IoT Arduino with Node MCU 8266 has had a significant impact on the monitoring and control of various systems. This technology has revolutionized the way data is collected and analyzed, offering a range of benefits for businesses and industries. With the integration of IoT Arduino and Node MCU 8266, organizations can now monitor and control their devices and equipment remotely, eliminating the need for manual intervention and reducing the risk of human error. This has resulted in improved efficiency and productivity, as well as cost savings.

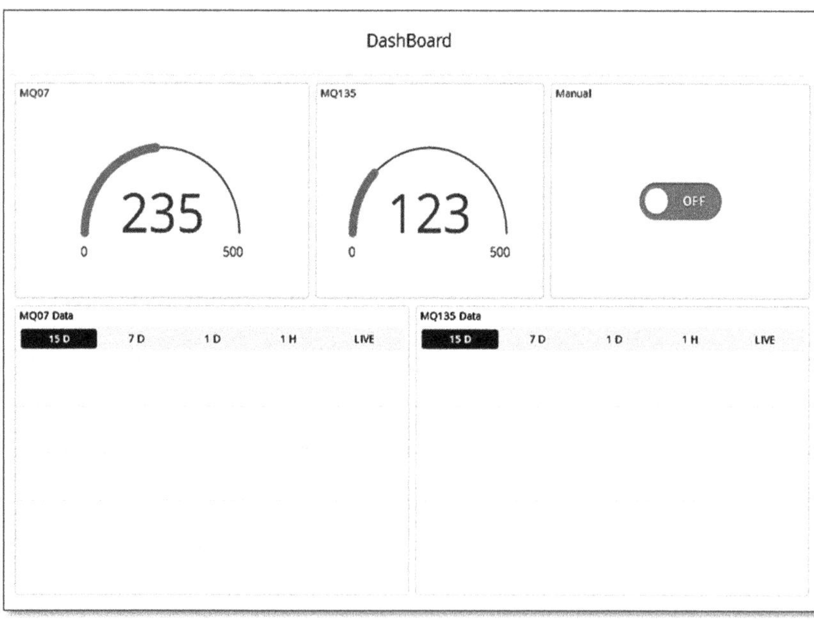

FIGURE 6.4 Dashboard data for air monitoring.

6.4.1.3 CODE ON IoT ARDUINO CLOUD WITH NODE MCU 8266 FOR AIR MONITORING

```
#include "thingProperties.h"
#include <Wire.h>
// Include any libraries required for the air quality sensor
// #include <AirQualitySensorLibrary.h>
void setup() {
Serial.begin(9600);
// Initialize the sensor and any other necessary setup
// Initialize IoT Cloud connection
initProperties();
// Connect to Wi-Fi
connectWiFi();
}
void loop() {
// Read sensor data
float airQualityValue = 0.0; // Replace this with your sensor reading logic
// Update the air quality property in IoT Cloud
airQuality = airQualityValue;
Serial.print("Air Quality: ");
Serial.println(airQualityValue);
ArduinoCloud.update();
delay(1000); // Adjust the delay based on your sampling frequency
}
```

6.4.2 FUEL MONITORING SYSTEM USE OF ESP8266 NODE MCU AND UBIDOTS

6.4.2.1 THE OBJECTIVE OF IoT FUEL MONITORING

1. **Real-time fuel consumption monitoring:** Tracking fuel usage is one of the goals of utilizing IoT in fleet management.

Fleet managers can obtain information on fuel levels, usage trends, and efficiency by incorporating IoT sensors and devices into cars. These data minimize expenses and optimize fuel usage.

2. **Preventive maintenance:** Enabling proactive and preventive maintenance is another goal of integrating IoT for fleet management. Fleet managers can get notifications about the condition of their vehicles and possible problems with them thanks to connected sensors and data analytics.

6.4.2.2 USE OF ESP8266 NODE MCU AND UBIDOTS

The ESP8266 Node MCU emerges as a robust solution for real-time fuel consumption monitoring due to its adaptability and seamless integration with various sensors, including both flow meters and ultrasonic sensors. This integration allows for precise fuel usage tracking, making it an optimal choice for applications requiring accurate data collection to manage fuel efficiently. Its compact size, low power consumption, and wireless connectivity extend its usability across diverse environments, from fleets and industrial machinery to remote locations. The Node MCU's compatibility with popular programming languages facilitates customization and enhancement of functionalities, enabling effective measurement and tracking of fuel levels. It efficiently transmits data to central systems or the cloud for analysis, finding practical applications in fleet management, industrial monitoring, and even smart home fuel tank systems. Its GPIO pins enable seamless interfacing with fuel sensors, ensuring continuous data acquisition for well-informed decision-making. Ultimately, this technology streamlines fuel usage, leading to improved operational efficiency through real-time monitoring and analysis.

By combining the ESP8266 Node MCU with flow meters and ultrasonic sensors, industries can benefit from improved operational efficiency through real-time monitoring and analysis of fuel consumption. The ESP8266 Node MCU's versatility allows for seamless integration of these specialized devices, enabling accurate measurement and quantification of fluid or gas flow within systems. This makes the ESP8266 Node MCU a valuable asset in industries such as transportation, manufacturing, and energy where precise fuel consumption monitoring is crucial for cost optimization and sustainability (Figure 6.5).

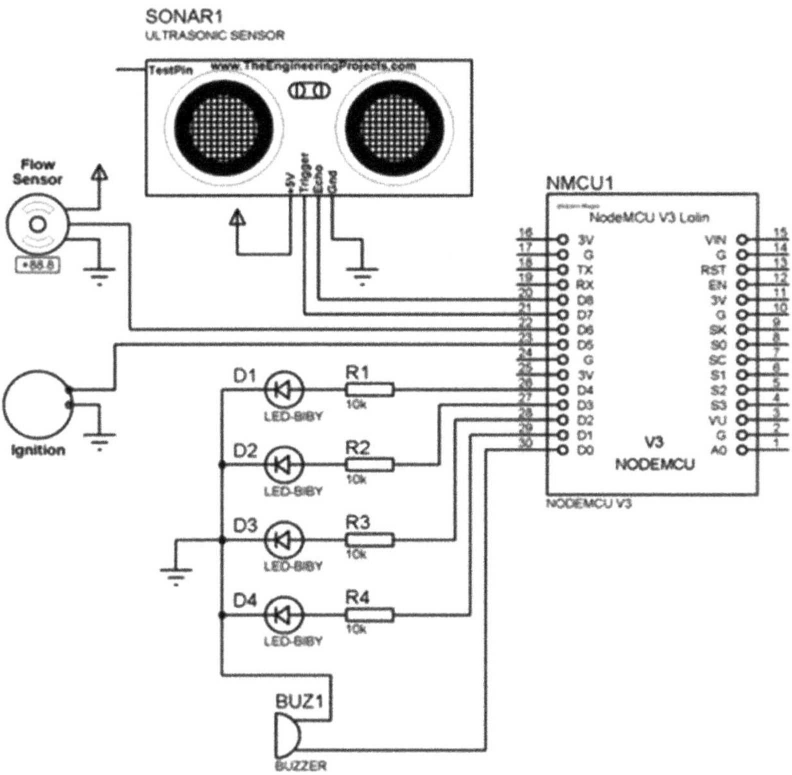

FIGURE 6.5 Fuel monitoring system use of ESP8266 Node MCU.

Users must attach a fuel level sensor (such as an ultrasonic sensor or a fuel level sensor) to the Node MCU board and transfer the data gathered to Ubidots to set up fuel monitoring using Node MCU ESP8266 and Ubidots. Figure 6.6 shows dashboard data for fuel monitoring is simple to gather and examine. These data contain details about fuel efficiency, usage, and any anomalies in fuel usage. Organizations can spot any problems or inefficiencies in their fuel management procedures by routinely reviewing this data. This enables them to adopt ways to optimize fuel consumption, cut expenses, minimize environmental impact, and make data-driven decisions.

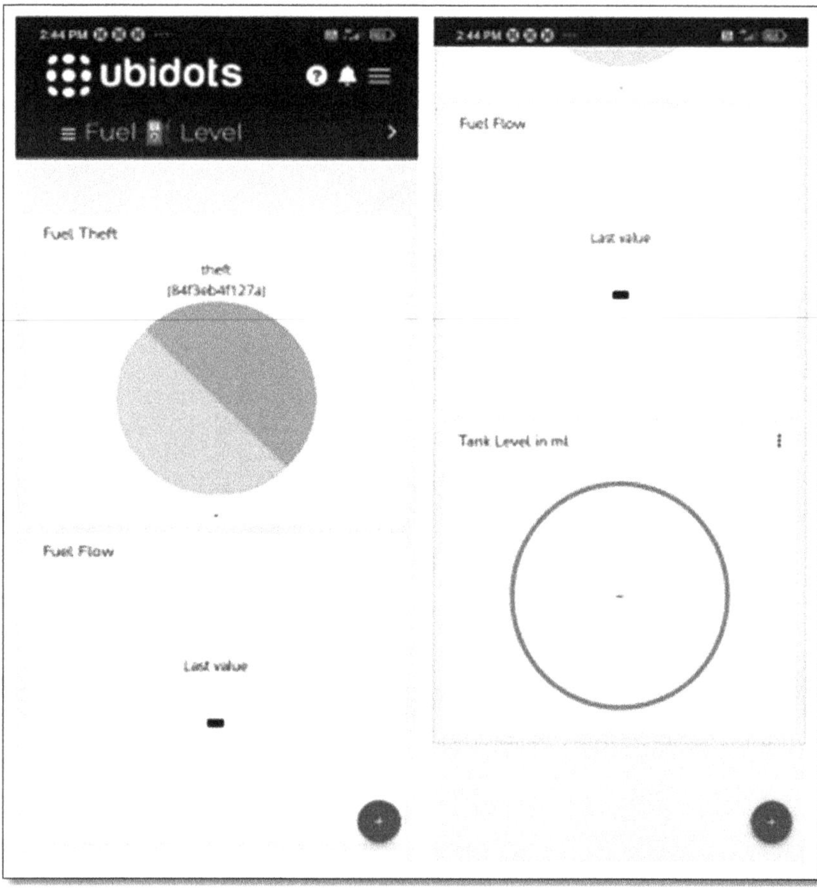

FIGURE 6.6 Dashboard data for fuel monitoring.

6.4.2.3 CODE ON UBIDOTS WITH NODE MCU 8266 FOR FUEL MONITORING

```
#include <ESP8266WiFi.h>
#include <Ubidots.h>
#define WIFI_SSID "YourWiFiSSID"
#define WIFI_PASS "YourWiFiPassword"
#define TOKEN "YourUbidotsToken"
#define DEVICE_LABEL "YourDeviceID"
```

```
Ubidots ubidots(TOKEN);
void setup() {
Serial.begin(115200);
// Connect to Wi-Fi
WiFi.begin(WIFI_SSID, WIFI_PASS);
while (WiFi.status() != WL_CONNECTED) {
delay(1000);
Serial.println("Connecting to WiFi...");
}
Serial.println("Connected to WiFi");
// Initialize the Ubidots client
ubidots.setDatasourceName(DEVICE_LABEL);
}
void loop() {
// Read fuel level from the sensor (replace with your sensor reading logic)
float fuelLevel = 0.0; // Replace this with your sensor reading logic
// Send fuel level data to Ubidots
ubidots.add(DEVICE_LABEL, fuelLevel);
bool bufferSent = false;
bufferSent = ubidots.send(); // Try to send the data to Ubidots server
if (bufferSent) {
Serial.println("Value sent to Ubidots");
} else {
Serial.println("Failed to send value to Ubidots");
}
delay(10000); // Adjust the delay based on your sampling frequency
}
```

6.5 CONCLUSION

The integration of IoT technology within commercial fleet sustainability initiatives is pivotal in monitoring and enhancing environmental impact. By harnessing IoT platforms like Arduino Cloud and Ubidots with NodeMCU ESP8266, comprehensive environmental monitoring is achievable. Fleet coordinators are better able to identify inefficiencies and implement the necessary changes to improve fuel efficiency by using real-time fuel consumption data from IoT-connected tracking devices. This reduces costs and improves the operation's sustainability and environmental friendliness. Additionally, the ability to track and monitor trucks in real time allows fleet managers to optimize routes and quickly alter them in reaction to traffic conditions or unforeseen events, ensuring on-time delivery and customer satisfaction. Cost is just one benefit of using IoT-connected tracking devices in fleet management, among many others.

First, integrating IoT Arduino Cloud with Node MCU 8266 facilitates precise air monitoring capabilities. Through this collaboration, real-time tracking and analysis of air quality metrics, such as pollutant levels, become streamlined. This offers fleet managers valuable insights into environmental conditions, aiding in proactive measures for air quality improvement. Second, the fusion of Ubidots with Node MCU 8266 for fuel monitoring is instrumental in bolstering sustainability efforts. This partnership enables the continuous monitoring of fuel consumption levels within the fleet. Accurate data collection and transmission to Ubidots empower decision-makers with the means to optimize routes and vehicle performance, curbing fuel wastage and reducing carbon emissions.

Both integrations underscore the transformative potential of IoT in advancing commercial fleet sustainability. The synergy between these technologies empowers fleet managers with actionable data, fostering informed decisions to mitigate environmental impact. Through these IoT-driven monitoring solutions, the pursuit of environmentally conscious practices in commercial fleet operations takes a significant stride forward.

KEYWORDS

- **commercial fleet sustainability**
- **IoT tracking**
- **environmental monitoring**
- **fleet management**
- **sustainability tracking**
- **green fleet**

- **eco-friendly transportation**
- **environmental progress monitoring**
- **IoT solutions for fleet sustainability**
- **sustainable logistics**
- **carbon footprint tracking**

REFERENCES

1. Zivelonghi, A., & Giuseppi, A. Smart healthy schools: IoT-enabled concept for multi-room dynamic air quality control. *Internet of Things and Cyber-Physical Systems* **2023**, *4*, 24–31.
2. Kumar, P., Martani, C., Morawska, L., Norford, L., Choudhary, R., Bell, M., & Leach, M. Indoor air quality and energy management through real-time sensing in commercial buildings. *Energy Build* **2016**, *111,* 145–153.
3. Dhanalakshmi, S., Poongothai, M., & Sharma, K. IoT based indoor air quality and smart energy management for HVAC system. *Procedia Computer Science* **2020**, *171,* 1800–1809.
4. Saini, J., Dutta, M., & Marques, G. A comprehensive review on indoor air quality monitoring systems for enhanced public health. *Sustainable Environmental Research* **2020** *30*, 6.
5. Prasad, R. V., Baig, M. Z., Mishra, R. K., Desai, U. B. & Merchant, S. N. Real time wireless air pollution monitoring system. *ICTACT Journal on Communication Technology* **2011**, *2*(2), 370–375.
6. Rahim, H. A., Dalimi, A., & Jaafar, H. Detecting drowsy driver using pulse sensor. *Jurnal Teknologi* **2019**, *73*(3).
7. Patil, S. In-vehicle driver health monitoring system. *International Journal of Technology and Science* **2019**, *9*(2), 38–40.

CHAPTER 7

Need-Based Irrigation Powered by IoT Toward Green Horticulture Cultivation

REENA CHANDEL[1,*] and DUSHYANT KUMAR SINGH[2]

[1]*Rayat Bahra University, Mohali, Punjab, India*

[2]*Edunet Foundation, Gurgaon, Haryana, India*

*Corresponding author

ABSTRACT

Groundwater levels are lowering as a result of India's over-reliance on irrigation for agriculture, making water shortage a serious problem. More than 90% of the nation's groundwater is used by the agriculture industry, which makes water management and irrigation efficiency more difficult. Traditional surface irrigation techniques are no longer sustainable as the demand for water rises and resources become more limited. In this regard, horticulture—which depends heavily on water resources to cultivate fruits and vegetables—offers a feasible way to increase water usage efficiency through the implementation of micro-irrigation and Internet of Things (IoT)-based automated systems. This chapter explores the preparation and development of an IoT-powered irrigation system that uses temperature and moisture sensors to provide real-time data to automate watering procedures. Through the provision of precise irrigation when required, the system seeks to improve agricultural water management and minimize water waste. The integration of IoT and smart sensors contributes to sustainable farming by reducing water consumption, enhancing crop yield, and improving farmers' financial outcomes.

IoT Potential for Green Energy Solutions. Abhishek Kumar, Ajay Roy, & Shamimul Qamar (Eds.)

DOI: 10.1201/9781779643599-7

7.1 INTRODUCTION

In India, precipitation varies greatly in both time and space, ranging from 11,000 to 90 mm. Water availability per person has fallen from 5000 m^3 in 1950 to 1545 m^3 in 2011 and is expected to fall even more in 2025 and 2050 to 1341 and 1140 m^3, respectively. The agriculture business alone utilizes more than 90% of all groundwater extracted for irrigation, accounting for 54.6% of total employment. Groundwater has grown as the primary source of irrigation over time due to its independence from other sources of water and reliable water supply. As a result of increasing reliance on groundwater, 64% of the country's districts witnessed water table reduction between TE2002 and TE2016. Utilizing irrigation potential, which includes both surface and groundwater, has expanded to 87 Mha, thanks to the combined efforts of government at all levels, while total irrigation potential has reached 140 Mha. There is a need to correct water use adeptness given the swiftly lessening available water. Microirrigation is an adept technique that may be critical in directing irrigation water demand. In contrast to surface watering procedures, which have a watering efficiency of only about 40%, correctly projected and managed trickle and watering systems for lawn watering orders have irrigation effectiveness of about 90% and 70%, individually. As a result of the government's acknowledgment of the gravity of the water shortage in many regions of the nation, calculating watering has enhanced a fundamental component of its water procedure. The administration combined all active watering programs into the Pradhan *Mantri Krishi Sinchayee Yojana* (PMKSY) in 2015, which involves data processing machine irrigation as a key component. The mechanic's skill and approachability of water-saving electronics through uniform support networks are two conditions for their two favorable deployments. The administration has started abundant programs to broaden the allure use of active watering electronics in agriculture. Depletion of the aquifer, unbalanced precipitation distribution, weak watering adeptness, tremendous and increasing break betwixt generated and utilized watering potential, frequent drynesses, and groundwater becoming worn are a few of the major challenges in watering [1].

A low-cost autonomous watering system will help struggling farmers save time and money. Agriculture is the lifeblood of India and many other countries. Agriculture in India is heavily reliant on rainfall and subterranean water for irrigation. Every day, subsurface water is rapidly decreasing. Over-irrigating a field may result in ineffective water management. Furthermore, the times of electricity distribution in fields are unusual. Farmers are compelled to sleep through several nights as a result. To manage these difficulties efficiently and with fewer human resources, modern technologies such as sensing and Internet of Things (IoT) must be deployed. One of the IoT's applications is a cutting-edge watering system. To alleviate difficulties, this essay tries to develop a method that automates watering while remaining manual. The recommended system was built using open-source electronics such as the Arduino Uno, as well as sensors such as soil moisture and temperature sensors. With the system's user-friendly interface, farmers can effortlessly run the irrigation process. The recommended method is cost-effective, given the cheap cost of the implementation hardware. Furthermore, it does not necessitate any additional labor costs [2]. In agriculture, technology is essential for successfully using irrigation resources, particularly subterranean water, providing support to farmers, increasing irrigation efficiency, putting a stop to quickly diminishing subsurface water, and improving farmers' financial status. The essay examines several automated irrigation systems and proposes an IoT-driven irrigation system for horticultural cultivation.

7.2 LITERATURE SURVEY

Irrigation has a greater impact on increasing the production and productivity of horticultural crops than any other single factor. Multiple irrigation system techniques are employed for various horticultural crops. Irrigation, in addition to sopping up the crops, can be critical in controlling insect pests in many horticultural crops. Mild irrigation helps to minimize termite infestations. Drip irrigation, as opposed to flood irrigation, keeps the frequency of tomato fruit borer in fields lower than sprinkler irrigation. Drip irrigation also aids in the control of insect pests in perennial orchards. To demonstrate this idea, the study evaluates and presents examples of ways to reduce the threat of insect pests in horticultural crops [3].

As to the United Nations, which has a significant agenda on the matter, up to 2 billion people are believed to be adversely affected by protein malnutrition and micronutrient deficiencies (vitamins and minerals). In response, the United Nations adopted the Sustainable Development Goals (SDGs) 2 and 12, which aim to end hunger and advance sustainable agriculture by boosting the production and consumption of fruits and vegetables. The relevance of Industry 4.0 technologies for resolving the full spectrum of horticulture concerns, from production to marketing in the context of sustainability, was not highlighted in previous studies, which only focused on the numerous horticultural challenges as they related to certain industries. The importance and uses of Industry 4.0 sciences for gardening are investigated in this place paper, containing the IoT, cloud estimating, artificial intelligence (AI), blockchain, and generous dossier. These technologies aim to help current procedures for affliction discovery, watering administration, fertilizer management, adulthood laborers, shopping, supply chain, soil virility, and weather patterns at pre-harvest, harvest, and post-harvest. The chapter studies difficulties and creates any important advice for future incident-established analysis. The study forms main approvals for relating to plant scenes, containing the use of robotics, drones accompanying dream science and AI for the discovery of contagions, weeds, plant diseases, and hunger, and edge-estimating handy instruments that may be grown with IoT and AI for forecasting and judging crop afflictions. Figure 7.1 mentions the duty of IoT in gardening [4–7].

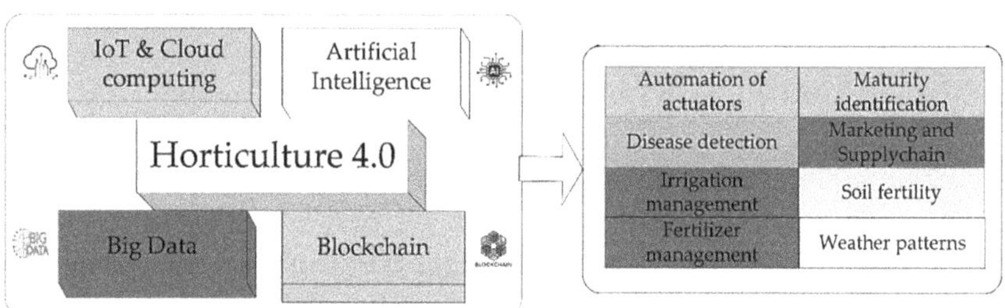

FIGURE 7.1 Role of IoT in horticulture.

A small and straightforward automated irrigation system is designed to utilize water and energy efficiently following plant requirements in coastal sandy soil. In the investigation, irrigation actuators, control processing units, and timing sensors with the model numbers RTC-DS1302, Arduino Mega-2560, and AC or DC pumps, respectively, were used. The mechanism also functions to automatically replenish the reservoir with water. For the production of onions (*Allium cepa* L.) and cabbage flowers (*Brassica oleracea* var. *botrytis* L.), automated watering of drip and sprinkler was used, powered by the solar system, in the coastal area of Banjarsari Village,

Cilacap, Central Java, Indonesia. As a result, the system that was used was able to continually deliver water to the plants at the adjusted times (07:00, 11:00, and 17:00, for 15 min each) [7].

Most Indians depend on farming to meet their often necessities, and it has a large impact on the frugality of the country. In dry domains or when skilled is the slightest precipitation, watering enhances troublesome. It must be enforced on laborers in consideration of highest in rank consequences and insulate the ruling class. Horticulture, a subsector of farming, donates significantly to the frugality, human food, neuter mainstreaming, and business. Horticultural fruit involves belongings like products, vegetables, flowers, spices, and sauces. They have of age firmly and immediately give reason for a large share of the land business. The mechanical irrigation control plan's objective search out weakens the exertions of the human controller (horticulturist) in gardening movements. The gardening system's purpose search out claim meal freedom measures. The study interprets an IoT-located soil dampness monitor that uses ESP32 accompanying an integrated Wi-Fi piece to monitor soil dampness level and please dossier from the board to the consumer's smartphone or desktop computer. Soil dampness monitors determine the amount of water in the soil and alert the farmer if the liquid sensor's fixed opening rate rises or falls, displaying overwatering or underwatering. This control whole is erected about an Arduino that was registered in embedded C. During this arrangement, the dossier from the sensor is transmitted, first to the cloud and before to the table attendant. The projected soil moisture monitor's judgment utilizing ThingSpeak guarantees that water is not emaciated [8].

During the 2014–2015 increasing seasons, a soil sensor-based robotic dampening system was proven in a marketing ornamental playroom. When distinguished from grower-governed watering, adoption of the sensor-located method resulted in an approximate 50% decline in watering requests (volume) in two ages. When comparing sensor-located watering to grower-trained watering in the production of *Pieris japonica*, *Hydrangea quercifolia*, and *Kalmia latifolia*, no alternatives in development or crop losses were noticed. When sensor-located irrigation was distinguished from raiser-managed watering in 2014, *Rhododendron catawbiense* displayed equal marquee size and lower death. However, in 2015, sensor-located watering control caused insignificant (>50%) Rhododendron losses. The use of averaged (across crop class) soil liquid values to demonstrate watering set points was deemed to have happened in extreme mortality.

Rhododendron umbrella building hindered water uptake in addition to the different three species on account of umbrella architecture. The alliance concerning this effect and precise watering inputs happened in chronic dryness environments within the Rhododendron block. Soil liquid sensor-located automatic watering may be an effective method of watering mechanization. Crop consultants' help is specifically desirable to minimize upset and maximize adoption all the while exercising. The use of soil moisture sensors to mechanize watering in commercial nurseries and greenhouses has proved to be an effective method of threatening water use in both cautiously regulated research studies and commercial trade acceptance trials [9, 10].

7.3 DESIGN AND DEVELOPMENT

A rapidly declining underground water table combined with ineffective irrigation techniques that result in over- and under-irrigation created a need for technical solutions to increase irrigation efficiency and, in turn, agricultural yield, improving farmers' financial standing and having less detrimental effects on the soil.

The use of technology in horticulture is examined in Ref. [7]. The IoT is mostly used in horticulture to monitor various pests and diseases that harm plants and have a severe negative impact on the industry. Moreover, fruits are detected by the integration of three basic item characteristics—color, shape, and feel—using IoT and cloud computing. Robots, drones with vision technology AI for identifying pests, weeds, plant diseases, and malnutrition, and edge-computing portable devices with IoT and AI for predicting and estimating crop disease were among the important recommendations for future research that the report covered. Even in the ideas for future development, irrigation is not included as an IoT use in horticulture. The general health of the crop is impacted by irrigation.

The IoT-driven irrigation system for horticulture (see Figure 7.2) is built around an Arduino UNO, an open-source processing unit, and a sophisticated soil volumetric moisture content (VMC) sensor. The device also detects light and humidity, which aids in accurate watering control. The farmer uses Telegram IoT to monitor the temperature, humidity, and soil VMC.

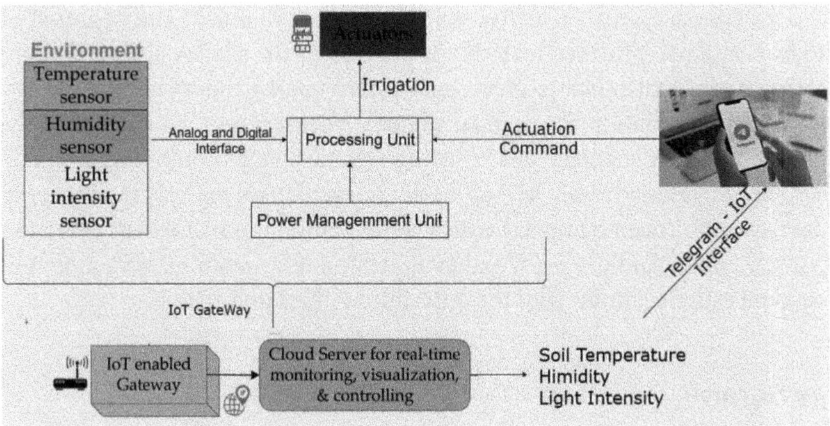

FIGURE 7.2 IoT enables irrigation monitoring and control for horticulture cultivation.

The soil VMC and air parameters were monitored by the automated watering system. An irrigation actuation instruction is issued if the measured parameters are found to be below the preprogramed threshold value for soil VMC. When the measured soil VMC reaches the desired value, the actuator initiates watering upon receiving the actuation signal. The processing unit then removes the irrigation actuation signal. If the farmer needs to start irrigation, they may also send an irrigation actuation instruction to Arduino using Telegram chat.

7.4 CONCLUSION

Growing fruits and vegetables are known as horticulture. By reducing malnutrition, it ensures production and consumption in the modern environment under the supervision of the UN. The accomplishment of the SDGs and digitization have been made possible recently by Industry 4.0 technology. Prior studies did not highlight the applicability and use of Industry 4.0 for certain horticulture problems. Owing to this constraint, the present investigation centered on the importance and utilization of Industry 4.0 technologies in horticulture, including the IoT, to enhance conventional watering techniques. Both carefully controlled academic studies and commercial company adoption trials have demonstrated that using soil moisture sensors to automate watering in commercial nurseries and greenhouses is an effective way to reduce water use. The adoption of sensors and IoT-powered technology progresses the society and agriculture sector toward sustainable green agriculture saving the water and energy utilized for irrigation for running pumps. According to the current study, further research should focus on using robotics, edge-computing portable devices developed with IoT and AI for predicting and estimating crop disease and drones equipped with vision technology and AI for detecting pests, weeds, plant diseases, and malnutrition. Horticultural irrigation will become more accurate with the application of AI and IoT.

KEYWORDS

- **irrigation**
- **Internet of Things**
- **soil moisture**
- **sensors**
- **soil volumetric moisture**

REFERENCES

1. Jain, R., Kishore, P., & Singh, D. K. Irrigation in India: Status, challenges and options. *Journal of Soil and Water Conservation* **2019,** *18*(4), 354–363.
2. Guntur, J., Raju, S. S., Jayadeepthi, K., & Sravani, C. H. An automatic irrigation system using IOT devices. *Materials Today: Proceedings* **2022**, *68*, 2233–2238.
3. Halder, J., Srivastava, K., & Rai, A. B. Irrigation, a potential tool for insect pest management in horticultural crops—a review. *Current Horticulture* **2023,** *11*(1), 3–7.
4. Singh, R., Singh, R., Gehlot, A., Akram, S. V., Priyadarshi, N., & Twala, B. Horticulture 4.0: Adoption of Industry 4.0 technologies in horticulture for meeting sustainable farming. *Applied Sciences* **2022**, *12*(24), 12557.
5. Sudarmaji, A., Sahirman, S., Saparso, & Ramadhani, Y. Time based automatic system of drip and sprinkler irrigation for horticulture cultivation on coastal area. In: *IOP Conference Series: Earth and Environmental Science*, **2019**, *250*, p. 012074.
6. Bhuiyan, M. N., Rahman, M. M., Billah, M. M., & Saha, D. Internet of things (IoT): A review of its enabling technologies in healthcare applications, standards protocols, security, and market opportunities. *IEEE Internet of Things Journal* **2021,** *8*(13), 10474–10498.

7. Čolaković, A., & Hadžialić, M. Internet of Things (IoT): A review of enabling technologies, challenges, and open research issues. *Computer Networks* **2018**, *144*, 17–39.
8. Nagaraj, P., Muneeswaran, V., Vasundhara, C., Koushik, K., Sumanth, B. S., & Naveen, P. Automated horticulture for farmers using IoT. In: *2023 2nd International Conference on Edge Computing and Applications (ICECAA) IEEE*, **2023**, pp. 1219–1222.
9. Wheeler, W. D., Chappell, M., van Iersel, M., & Thomas, P. Implementation of soil moisture sensor based automated irrigation in woody ornamental production. *Journal of Environmental Horticulture* **2020,** *38*(1), 1–7.
10. Van Iersel, M. W., Chappell, M., & Lea-Cox, J. D. Sensors for improved efficiency of irrigation in greenhouse and nursery production. *HortTechnology* **2013,** *23*(6), 735–746.

CHAPTER 8

Empowering Smart Farming: Extensive Plant Surveillance and Accurate Pest Identification

M. JOTHIBASU[1,*], P. V. MIRUTHULA[2], and K. SHREE HARINI[2]

[1]*Department of Electronics and Communication Engineering, PSG Institute of Technology and Applied Research, Coimbatore, India*

[2]*PSG Institute of Technology and Applied Research, Coimbatore, India*

*Corresponding author

ABSTRACT

Modern agriculture has been transformed by the Internet of Things (IoT), which provides creative ways to increase pest control, maximize resource use, and improve plant care. In order to track and maintain ideal conditions for plant growth, this chapter examines the integration of IoT in the development of a smart plant monitoring system that integrates many sensors. Gardeners and farmers struggle to maintain the right combination of light, temperature, moisture, and insect management in both urban and rural farming areas. In order to solve these problems, the suggested system integrates sensors for temperature, humidity, light intensity (LDR), and wetness with an Arduino Uno to enable automated monitoring and decision-making. In order to ensure that plants receive the proper amount of water and avoid both overwatering and under watering, the system includes an autonomous irrigation mechanism. To further improve plant care, color sensors are also being used to track the health of plants and insect detection technology. In addition to having immediate uses in small-scale contexts like rooftop gardens. The agricultural industry can solve

IoT Potential for Green Energy Solutions. Abhishek Kumar, Ajay Roy, & Shamimul Qamar (Eds.)

DOI: 10.1201/9781779643599-8

major issues like crop yield, water waste, and pest control by combining these technologies, opening the door to more intelligent and sustainable farming methods.

8.1 INTRODUCTION

In today's world, there are billions of useful devices connected to networking, and these will go on expanding over time. The Internet of Things (IoT) has played the most significant role in the physical world by linking unique devices with limited programming. It represents the set of information, sensors, software, and technology that are inter-networked in a distributed manner and efficiently process these data. In the present time, though people have a lot of desire to grow plants and have a garden at home, they tend to miss their daily schedule toward taking care of their plants. And the technology is quite with them to support with a reminder to the people. Also, each plant should be taken care of with the amount of light, temperature, and water they receive, depending on its needs in the right quantity with no deviations. Also, the monitoring of a plant on its health condition is far more necessary along with its needs. The objective of this chapter is to unite all the sensors and their functions to achieve a smart plant monitoring system with the help of the IoT. This proposed system will work based on the information from the sensors and enhance plant growth accordingly.

Plants, as living entities, play an indispensable role in our lives, serving as a vital link between humans and the natural world [1, 2]. Their significance in our existence on this planet cannot be overstated. Therefore, the development of a project that provides plants with an optimal environment and incorporates pest detection is of utmost importance. This innovative project revolves around the meticulous observation and continuous monitoring of various crucial parameters essential for plant growth and well-being. By employing a range of sensors, it effectively tracks and records the plant's temperature, humidity, moisture levels, pest control measures, amount of light received, and even the rainfall in its surroundings. These sensors gather valuable data, which is then transmitted to an Arduino Uno for processing and controlling the necessary functions (see Figure 8.1). One of the primary features of this project is an automatic irrigation system, which ensures that plants receive the appropriate amount of water automatically [3]. While this project finds immediate application in small-scale rooftop gardening, its potential extends far beyond that. With further development and refinement, it can be adapted for large-scale implementation in agricultural fields. This advancement holds immense promise for optimizing irrigation practices, reducing water wastage, and enhancing crop productivity on a broader scale.

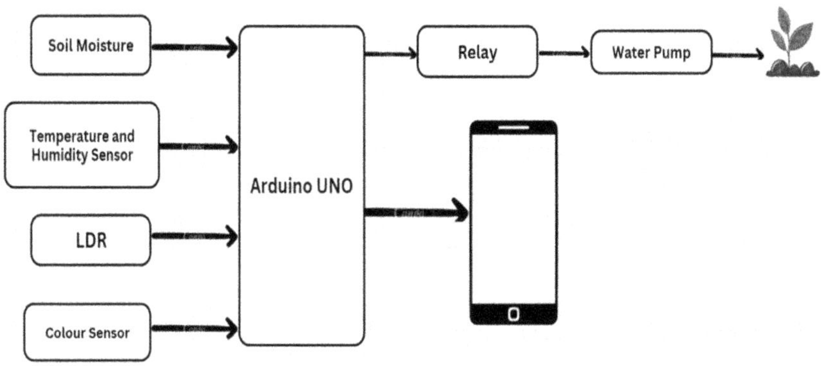

FIGURE 8.1 The flow of execution.

By providing a comprehensive solution that combines environmental monitoring, pest detection, and automated irrigation, this project aims to revolutionize plant care and cultivation practices. It empowers individuals, whether they are urban gardeners or professional farmers, to create thriving green spaces and improve agricultural efficiency. Ultimately, the integration of technology and nature in this manner fosters a sustainable and harmonious relationship between humans and plants, bolstering our coexistence on this planet.

8.2 RELATED WORKS

Pawar et al. proposed an IoT-based automatic watering system for agriculture. It uses sensors like moisture, humidity, and temperature sensors to monitor soil conditions and water levels in tanks. The data is sent to a web server, allowing remote monitoring and control through a mobile application. The system aims to reduce water wastage, automate irrigation, and improve farming efficiency, ultimately leading to more profit for farmers. Nawaz et al. [5] have proposed a system for agricultural disease and insect pest control using sensor nodes to monitor temperature, humidity, and leaf color to detect plant diseases. Future work includes integrating image processing for disease classification and improving parameter accuracy based on varying climatic conditions. Bhadani and Vashisht [9] developed the project for Indian farmers using Arduino Uno to monitor soil moisture, temperature, and humidity to enhance crop growth. It showcases the expansion of computer programming in agriculture and the potential for fully automated farming systems. Thakare and Bhagat [13] presented a paper on an economical smart irrigation system using Arduino and Wi-Fi modules for farm automation. It detects soil moisture, pH level, and temperature to regulate irrigation and sends notifications through an IoT platform for remote monitoring. The system offers efficiency, cost-effectiveness, and real-time updates, making agriculture smarter with automation and IoT technology. Rawal proposed an automated irrigation system using soil moisture sensors and IoT technology to maintain optimal soil moisture levels. The system sends updates to farmers via GSM-GPRS and provides real-time sensor readings on a webpage and ThingSpeak channel for analysis. By automating the irrigation process, this system saves time, conserves water, and improves farming efficiency, showcasing the potential of IoT and automation in agriculture. Rawidean et al. developed a system for the greenhouse. Monitored valves in the field were pH, temperature, and soil moisture valves mounted on different areas. Nisha and Megala [14] utilized soil moisture and temperature sensors, Zigbee protocol for data transmission, and a microcontroller for irrigation control. It is powered by solar panels and incorporates a cellular Internet interface for data inspection. Additionally, a wireless camera is employed for disease monitoring using image processing. This low-cost and energy-efficient system is particularly beneficial in water-limited and geographically isolated areas. The proposed system allows farmers to update and monitor crop field information through a PC or mobile device. Vaishali et al. [15] focused on parameters like temperature and soil moisture, presenting a mobile integrated and smart irrigation system using IoT for application-controlled monitoring. Considering the increasing demand for irrigation due to population growth, addressing water availability and implementing controlled irrigation through sensor-based systems has been considered an effective solution. The system automates watering based on ground moisture levels, benefiting farmers and gardeners who face time constraints. The system emphasizes power efficiency, cost-effectiveness, scalability, and ease of use, with future potential for outdoor utilization and further development. Junxiang and Haiqing developed the system which was a combination of embedded technology and a web server for the surveillance of greenhouse systems; they mainly concentrated on the software part of embedded technology. Liqiang et al. focused on the importance of IoT in promoting agricultural informatization and presented a crop-monitoring system based on wireless sensor networks. The system included two types of nodes for environmental parameter acquisition and image capture. The architecture and system software are described, and experimental results demonstrate the system's effectiveness. Future tasks included enhancing stability, adapting to complex environmental changes, and improving low-power consumption performance. Jaishetty et al. proposed the concept of IoT and its application in agriculture. It presented a novel methodology for physical parameter monitoring, data integration to the cloud, alert generation, and predictive analysis using MATLAB. The implementation includes sensors for temperature, rainfall, light, and moisture, with data visualized in real time. Message queuing telemetry transport protocol is utilized for machine-to-machine communication, enabling remote control of the agricultural field. The work emphasizes the importance of technological improvement in crop productivity and highlights the need for data mining and predictive analysis in agriculture. The proposed IoT architecture offers real-time data analysis and global collaboration for understanding crop behavior. The system exhibits optimal latency, high packet delivery rates, and accuracy, with the potential for further improvement through self-learning techniques in the cloud. Yakkundimath focused on detecting plant diseases by analyzing temperature, humidity, and color parameters of leaves. It proposes a three-level system that utilizes sensor nodes to collect data on diseases and pests, enabling automated detection and monitoring. The proposed approach can be applied by farmers, industry professionals, botanists, food engineers, and healthcare providers. Future work includes incorporating image processing techniques for improved accuracy and expanding the system's capabilities to identify specific diseases. Limitations include the current focus on leaf health detection and the need for more precise parameter values based on varying climatic conditions. Ibrahim Mat and Mohamed Rawidean MohdKassim et al. proposed a system for agriculture application that is irrigation to save water. The test shows savings of 1500 ml of water per day per tree. Results were observed graphically. Kumar et al. focus on a cost-effective smart irrigation system for middle-class farmers in India. The system utilizes a soil moisture sensor to automatically control the water motor and direct the flow of water in the field. The information about the operation of the motor and water direction is sent to the user's mobile as both a message and an email. This prototype demonstrates successful automatic control of irrigation based on feedback from the soil moisture sensor, leading to optimized water usage and reduced human intervention for farmers. Yewale et al. presented a research paper on a fully automatic object sorting system based on colors using the TCS3200 sensor and Arduino Uno board, providing cost-effectiveness and efficient results for process industries. Duraianand and Sivasangari have done research work that focuses on using IoT systems with temperature and color sensors to detect early-stage diseases in banana plants, contributing to improved crop protection. With a 92.66% accuracy rate, the low-cost IoT-based system can identify diseased leaves, potentially benefiting farmers and enhancing banana yield. Sayanthan et al. have done a study utilizing an Arduino-based soil moisture investigator to determine the crop water requirement for eggplant. By comparing the actual water requirement (149.26 L) with the water applied by farmers (604.8 L), it was found that farmers were wasting excess water (4.05 times higher). The proposed low-cost soil moisture sensing technique can automate irrigation and conserve water effectively. Following are the drawbacks of the existing system which the authors have learned through a literature survey.

1. Not developed for all kinds of farming.
2. Lack of user-friendly interface.

Unlike other projects, this system employs an extensive array of sensors to provide unparalleled and precise monitoring of diverse plant growth parameters. The utilization of an advanced color sensor to detect subtle color variations in plant leaves sets it apart. This innovative feature enables early pest infestation detection, contributing to preventive pest management.

8.3 PROPOSED WORKING MODEL

The circuit configuration encompasses a sophisticated array of integrated sensors meticulously designed to monitor and assess the overall growth of plants. This intricate system is effectively powered by an external energy source, usually in the form of a durable and reliable battery, to ensure uninterrupted and seamless operation. Situated close to a Mexican mint plant, scientifically recognized as *Coleus amboinicus*, the circuit intelligently captures and analyzes crucial environmental parameters including light intensity, temperature, and humidity levels. Moreover, it harmoniously incorporates an ingenious automated irrigation system, facilitated by a highly sensitive moisture sensor, to meticulously regulate and sustain the ideal moisture content within the soil, thereby optimizing plant health. An exceptional attribute of this system lies in its ability to accurately detect potential pest infestations through the utilization of a cutting-edge color sensor. By discerning the subtle variations within the range of vibrant green hues exhibited by the plant's leaves, any deviations or alterations in leaf coloration can be swiftly identified [4–7]. This crucial functionality empowers users to proactively identify and address pest-related issues or any other abnormalities that may arise, safeguarding the overall well-being and vitality of the plant [4–7]. In conclusion, the integration of these state-of-the-art components culminates in a highly comprehensive and sophisticated system, purposefully crafted to diligently monitor, nurture, and maintain the growth of plants.

The circuit diagram of the proposed model is shown in Figure 8.2. The colored lines represent the connection of the circuit elements. To connect various lines within a single port, the breadboard is used.

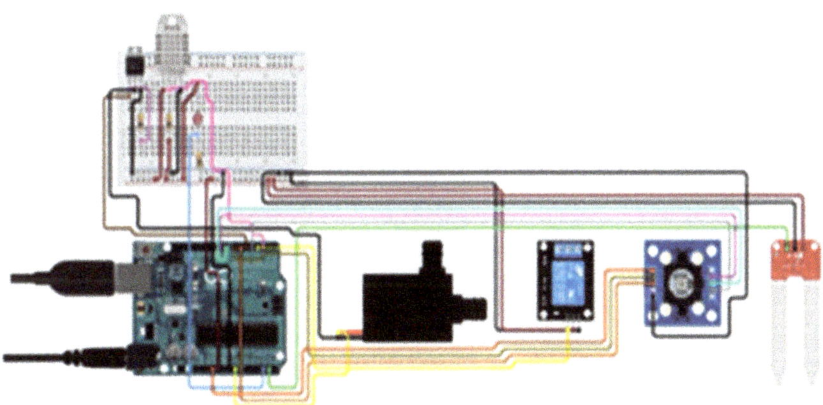

FIGURE 8.2 Circuit connections of the proposed.

In the circuit diagram, each sensor's output will be connected to the appropriate analog or digital input pins of the Arduino. The Arduino will process the sensor data, and based on the programmed logic, it will control the motor or other output devices. For example, if the moisture sensor detects low soil moisture, the Arduino may trigger the motor to activate a watering system. Similarly, temperature, humidity, and color sensors can be used to trigger specific actions or responses based on the measured environmental conditions or object colors.

The comprehensive working of the system is presented in the flowchart shown in Figure 8.3. The flowchart illustrates the case conditions for the sensors, enabling a thorough comprehension of how the system operates and responds based on the sensor readings [8].

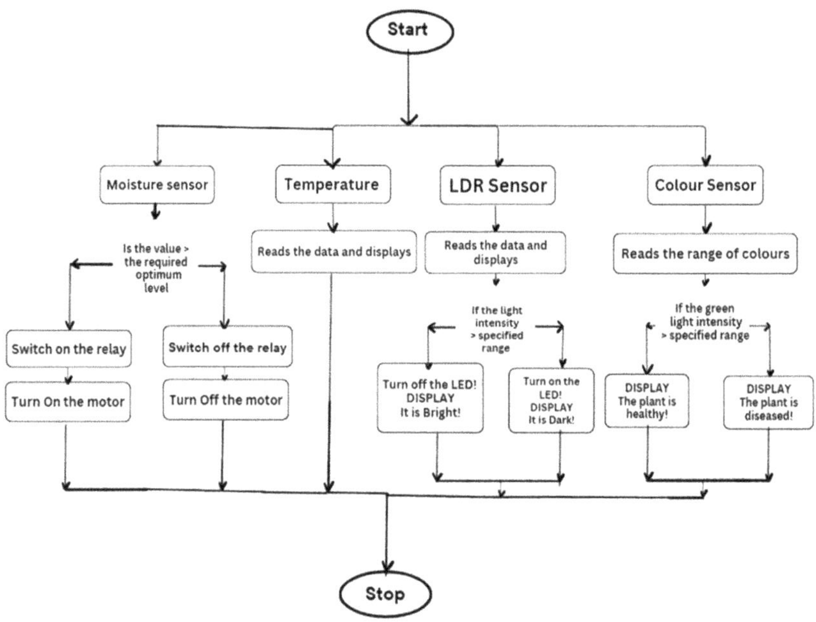

FIGURE 8.3 Sequential execution of code.

8.3.1 *FOR THE MOISTURE SENSOR [9–12]*

Step 1: Enter the main loop. Read the analog value from the soil sensor and store it in the variable "water."

Step 2: If the water level is less than or equal to the threshold, turn off the relay by setting the digital pin 13 to LOW. Print the water level and a message indicating that the relay has been cut.

Step 3: Otherwise, if the water level is above the threshold, keep the relay on by setting the digital pin 13 to HIGH. Print the water level and a message indicating that the relay continues to provide a signal and water supply.

8.3.2 FOR TEMPERATURE SENSOR [9]

Step 1: Read temperature and humidity from the DHT11 sensor, storing the values in variables.

Step 2: If the read operation fails, print the error details (error code and duration) and return.

Step 3: Print the sampled temperature and humidity values in the desired format.

8.3.3 FOR LDR SENSOR

Step 1: Read the LDR (Light Dependent Resistor) sensor's status and store it in the ldrStatus variable.

Step 2: If the ldrStatus is greater than or equal to the threshold, turn on the LED by setting the ledPin to HIGH. Print a message indicating that it is dark and also display the current ldrStatus value.

Step 3: Otherwise, if the ldrStatus is less than the threshold, turn off the LED by setting the ledPin to LOW. Print a message indicating that it is bright and display the current ldrStatus value.

8.3.4 FOR COLOR SENSOR

Step 1: Set the photodiodes to read the GREEN (G) frequency.

Step 2: Read the output frequency from the sensor and store it in the greenFrequency variable. Print the value of the GREEN (G) frequency.

Step 3: If the green frequency is greater than the threshold, print "It is diseased;" otherwise, print "It is healthy."

8.4 HARDWARE REQUIREMENTS

This system mainly uses an Arduino board as the development board; the list of components to build the model is listed below along with its perfect model number (if available).

1. Arduino Uno
2. Moisture Sensor
3. Temperature and Humidity Sensor—DHT-11
4. Rain Sensor
5. LDR Sensor
6. Color Sensor
7. 5 V Water Pump
8. 5 V Relay Module

8.4.1 ARDUINO UNO

Arduino Uno is a microcontroller board based on the ATmega328P. It is an interactive board that can be used to build projects and control various objects. It has 14 digital input/output pins, six analog inputs, a USB connection, a power jack, an ICSP header, and a reset button. It is designed to be user-friendly and easy to use, making it perfect for beginners and experienced makers alike. The Arduino Uno also comes with the Arduino IDE (Integrated Development Environment) which allows users to write code and upload it to the board. Additionally, it can be powered by an external 9-V battery or through a USB connection from your computer. It has 32 K bytes of flash memory and 2 K bytes of SRAM to store the program that is coded for operation (Figure 8.4).

8.4.2 MOISTURE SENSOR

Moisture sensors are designed to measure the volumetric content of water in a given material. These sensors usually have four pins and operate on a 5-V power supply. They use an electric dielectric constant to measure the moisture content of the material. The optimum temperature for these sensors is usually between 0°C and 50°C. It utilizes capacitance to gauge the water content of the soil (dielectric permittivity). Moisture sensors are used in various applications such as agriculture, construction, and industrial settings (Figure 8.5).

FIGURE 8.4 Arduino Uno.

FIGURE 8.5 Moisture sensor.

8.4.3 TEMPERATURE AND HUMIDITY SENSOR

It is a digital temperature and humidity sensor and uses a digital signal acquisition technique. This sensor has three pins: VCC, data pin, and ground pin. The data pin acts as output for both temperature and humidity through serial data. For humidity sensing, we have a capacitor with two electrodes present between the moisture-holding substrate, where the change in the capacitance value occurs with a change in humidity level. The temperature sensor is made up of semiconductor ceramics or polymers. For measuring the temperature, it has a thermistor to measure the surrounding air temperature. It has a negative temperature coefficient, and it reduces with the increase in the temperature. This sensor requires about 3–5 V and the maximum current is 2.5 mA. The humidity of 20–80% has 5% accuracy and for 0–50°C, it has about 2°C accuracy. It has a 1-Hz sampling rate for every second, and before sending information to Arduino, it needs some period delay; the entire procedure time is 4 min (Figure 8.6).

FIGURE 8.6 Temperature and humidity sensor—DHT11.

8.4.4 RAIN SENSOR

Rain sensors are electronic devices used to detect the amount of rainfall and measure its intensity. They are typically connected to a 5-V power source, with GND and D0 pins for data input and an A0 pin for analog output. Rain sensors usually consist of a nickel plate on the side, and sensitivity is adjusted using a potentiometer. The potentiometer is connected to an analog-to-digital converter which then records the data from the sensor. This data can then be used to determine the amount and intensity of rainfall in a given area. It is used in various applications such as irrigation systems, weather monitoring stations, and automatic car washes. They can also be used to alert people when it rains so that they can take necessary precautions (Figure 8.7).

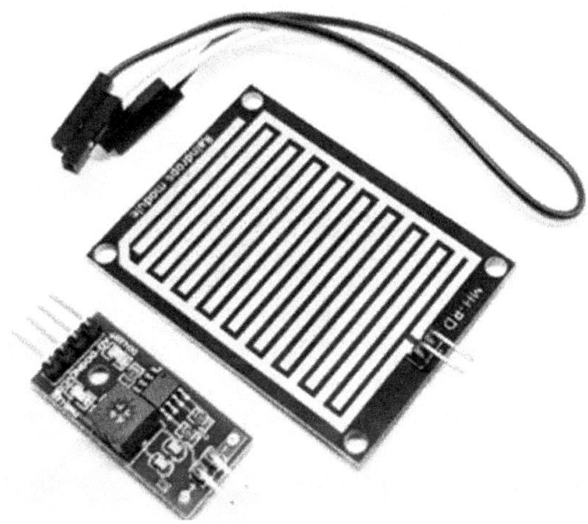

FIGURE 8.7 Rain sensor.

8.4.5 LDR SENSOR

LDR refers to the light-dependent resistor, also known as the photoresistor. It is a passive component and is used to measure the intensity of the light. This sensor uses the principle of photoconductivity, that is, when the light falls on the surface, the conductivity of the material reduces (photosensitive semiconductor materials like cadmium sulfides (CdS)), the electrons in the valence band are excited to the conduction band increasing the number of the charge carriers whenever the energy of the photon is greater than the energy of the band gap. The resistance increases with the reduction in light intensity and vice versa. Resistance in darkness will be about 1 MΩ and the brightness will be reduced up to a couple of kilo ohms (Figure 8.8). Here, the resistors are pure semiconductor devices like silicon or germanium. With the constant voltage, the LDR light intensity increases, and the current increases. The rated power of LDR is about 200 W with dimensions about 3–20 mm. It has a response time of 2–50 min. When we have sufficient light in its environment, that is, on surface of LDR, the converted digital values read from it through the Arduino will be in the range of 800–1023.

FIGURE 8.8 LDR sensor.

8.4.6 *COLOR SENSOR—TCS3200*

TCS3200 color sensor uses TAOS TCS3200 RGB sensor chip to detect the color, and appropriate action can be triggered. It has eight pins having four selection pins as input and an enable for output frequency (active low) along with a ground pin and voltage supply pin. The four LED lights present light up the object in front of it. It contains an 8×8 array of photodiodes with four different filters, that is, 16 photodiodes in each red, green, and blue, sensitive to respective wavelengths, with an additional 16 photodiodes without any filter. It requires about 2.7–5.5 V having a digital TTL interface. The sensor has a current-to-frequency converter that converts the photodiodes' readings into a square wave with a frequency that is proportional to the light intensity of the chosen color. Then, this frequency is read by Arduino Uno. It communicates directly to the microcontroller and produces full-scale output frequency (Figure 8.9 and Table 8.1).

FIGURE 8.9 Color sensor—TCS3200.

TABLE 8.1 RGB Color Sensor Using the TCS3200 Module

S0	S1	Output Frequency Scaling (f_0)	S2	S3	Photodiode Type
L	L	Power down	L	L	Red
L	H	2%	L	H	Blue
H	L	20%	H	L	Clear (no filter)
H	H	100%	H	H	Green

8.4.7 WATER PUMP

This submersible water pump is designed to operate within a 3–6 V power range, making it versatile for various applications. It boasts a low-power consumption of only 220 mA, ensuring energy efficiency. This pump can efficiently move up to 120 L of water per hour, suitable for tasks like water circulation or transfer. Its innovative water suction mechanism draws approximately 70 L of water through the inlet and expels it through the outlet, making it a valuable tool for both small-scale and low-power water management needs (Figure 8.10).

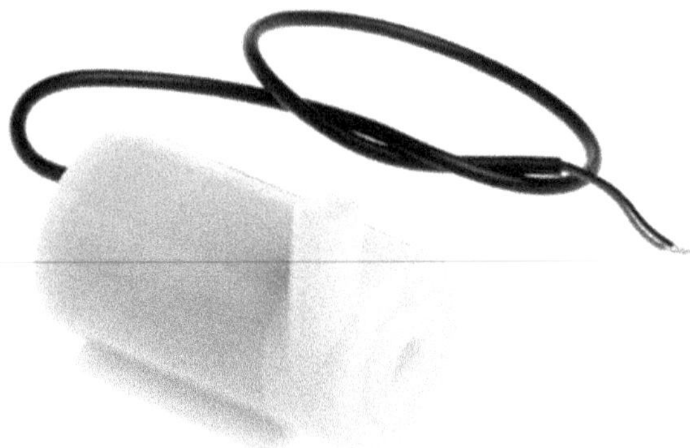

FIGURE 8.10 Water pump.

8.4.8 RELAY MODULE

A relay module is a magnetically operated switch that is used to control high voltage or current circuits using a low voltage signal. It works by inducing a magnetic field on a metal strip, which then triggers the switch when the positive signal from an Arduino board is applied. There are three external pins available in the relay module to connect the external circuit as shown in Figure 8.11. It can be used to control various types of electrical devices such as motors, lights, and fans. It enables users to save time and energy by eliminating the need for manual switching.

8.4.9 NRF24L01 WIRELESS TRANSCEIVER MODULE

The NRF24L01 module is a highly versatile communication module widely used in various applications, especially in the IoT domain. Operating within the 2.4 GHz ISM band, it enables reliable wireless communication over a range of several hundred meters. With support for multiple channels and data rates of up to 2 Mbps, it offers flexibility in transmitting data based on specific requirements. One of its key features is a robust transmission protocol that incorporates packet acknowledgment and automatic retransmission, ensuring consistent and error-free communication. Additionally, the NRF24L01 module integrates power-saving capabilities, optimizing energy consumption for efficient operation. In summary, the NRF24L01 module provides a reliable, flexible, and energy-efficient wireless communication solution, making it a valuable choice for various IoT applications (Figure 8.12).

FIGURE 8.11 5-V relay module.

FIGURE 8.12 NRF24L01 module.

8.4.10 POWER SOURCE

The power source for the working of the Arduino is powered via an external power supply or with a USB connection. The USB connector usually provides a regulated 5 V to power the board electronics. The board also has an AC socket that can be used to power the boards for additional voltage supply when needed. The power supply adapter that provides from 7 to 12 V of DC is required.

8.5 EXPERIMENTAL RESULTS

A visual representation of the circuit connections, illustrating the wiring and interconnections of the components in the proposed system is provided in Figures 8.13 and 8.14. Figure 8.13 represents the various sensors that are connected to the Arduino, and the color sensor is placed close to the Mexican mint plant, scientifically recognized as *C. amboinicus* for the detection of the leaf color. In Figure 8.14, the moisture sensor is placed in the soil and connected to the relay and the water pump to activate the sufficient water supply.

FIGURE 8.13 Circuit connection—plant monitoring.

FIGURE 8.14 Circuit connection—irrigation system.

Figure 8.15 showcases a user-friendly application interface designed for easy access and efficient monitoring. This interface offers various monitoring options, each presenting sensor values and alert messages in real time. Developed using Flutter, the application seamlessly integrates with Bluetooth sensors, enabling the swift transfer of data. Users can effortlessly access critical information and receive timely alerts, empowering them to monitor the system effectively. This smart interface enhances the overall user experience by providing convenience and precision in system management. It fosters optimal decision-making for improved agricultural practices, allowing users to make informed choices that contribute to better plant care and resource utilization. In essence, Figure 8.15 represents a technological advancement that simplifies the monitoring process, making it more accessible and user-centric, thereby promoting efficient and effective agricultural management.

Figure 8.16 serves as a comprehensive dashboard, offering vital insights into environmental conditions for optimized plant care and resource conservation. The "sample ok" line presents temperature in Celsius and humidity levels (e.g., 72 H), providing users with immediate information about the plant's microclimate. This data helps users adjust conditions like ventilation and irrigation to ensure the plant's well-being.

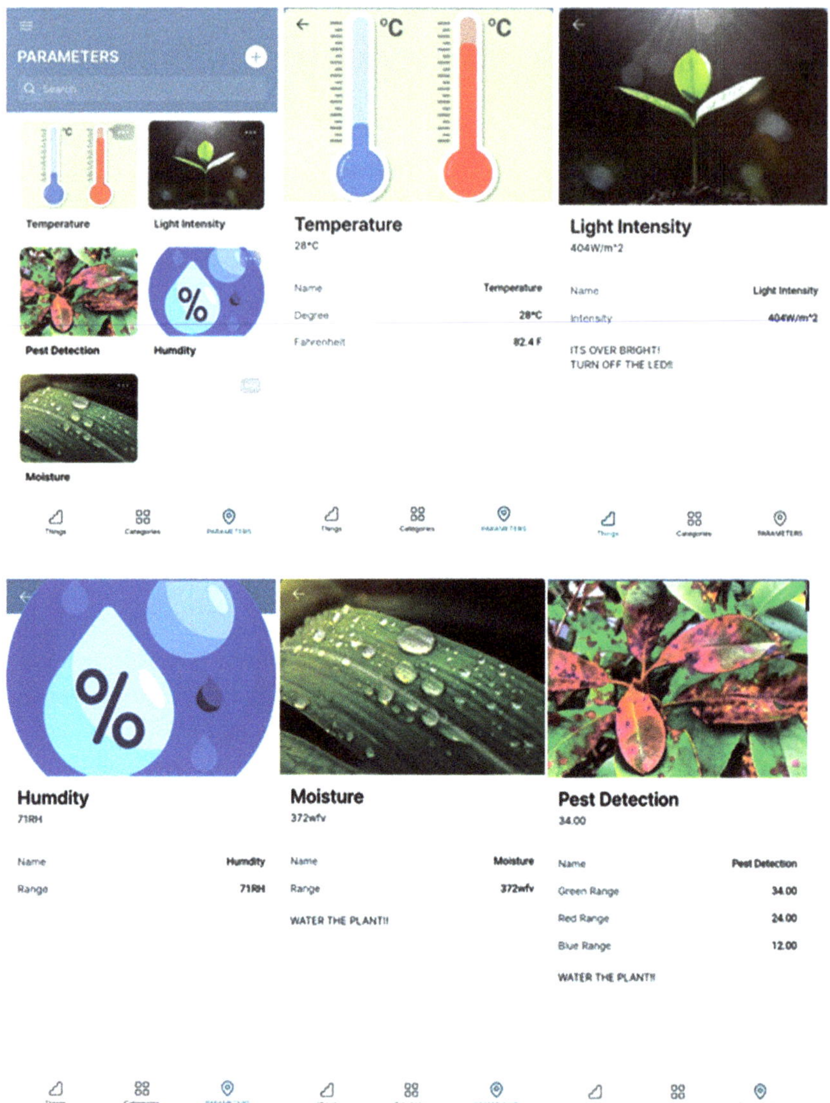

FIGURE 8.15 The application developed for Temperature: Environmental warmth, Humidity: Atmospheric moisture, Light intensity: Illumination levels, Soil moisture: Hydration status, and Pest detection.

The LDR sensor's value of 404 triggers the message "It is bright, turn off the LED." This intelligent response to ambient light conditions not only conserves energy but also supports natural light patterns, promoting plant health. The subsequent line displays color sensor values for red, green, and blue intensities, along with the message "It is healthy." This information reflects the balance of light spectrum and pigment development, ensuring the plant receives suitable light for photosynthesis. The digital value "wet" from the rain sensor and moisture sensor readings (e.g., 390) prompt messages like "Stop watering" or "Water the plant," enabling adaptive watering control. This prevents over- or under-watering, conserving resources, and promoting plant growth.

```
Sample DHT11...
Sample OK: 28 *C, 72 H
Its BRIGHT, Turn off the LED : 404
R = 24 G = 34 B = 12
It is HealthyDigital value : wet


390
stop watering
391
stop watering
373
water the plant
373
water the plant
```

FIGURE 8.16 Sensor recording.

8.6 CONCLUSIONS

The integration of IoT technology and digital advancements in agriculture has brought about a profound transformation in both rural and urban areas. The connectivity of devices and the collection of data from sensors have revolutionized the way farmers and agricultural practitioners manage their resources, ultimately leading to increased productivity. This transformative process has been further streamlined by the development of mobile applications that are accessible from anywhere with an Internet connection, offering a cost-effective alternative to larger and more expensive instruments. One of the most notable aspects of this technological revolution is its ability to effectively tackle some of the major challenges in agriculture. By harnessing IoT technology, the agricultural sector can avail itself of practical and efficient solutions. However, as with any technological innovation, there are always opportunities for further enhancement and expansion. The future of IoT in agriculture holds immense potential for additional advancements and innovation.

Sensor technologies have been critical in the success of IoT in agriculture. These sensors can monitor various aspects of the farming environment, such as soil moisture, temperature, humidity, and even the health of livestock. Future advancements in sensor technology can lead to even more accurate and real-time data collection. For instance, sensors that can detect the presence of pests or diseases in crops can provide early warnings, enabling farmers to take proactive measures to protect their yields. In addition to improving sensor technology, enhancing data analytics algorithms is another area ripe for development. Currently, data collected from IoT devices can be overwhelming, and making sense of this data is a challenge. With advanced algorithms, it becomes possible to process and analyze this data efficiently, providing actionable insights to farmers. Machine learning and artificial intelligence hold the key to making IoT applications in agriculture even more powerful. The future of IoT in agriculture also offers tremendous potential for innovation in decision-making tools. Advanced decision support systems can help farmers make informed choices regarding crop selection, pest control, and resource allocation. These tools can integrate data from various sources, including IoT devices, weather forecasts, and market trends, to provide comprehensive recommendations for farmers.

In conclusion, the integration of IoT technology and digital advancements in agriculture has ushered in a new era of innovation and efficiency in the sector. With continuous improvements in existing applications and the exploration of new possibilities, the future of IoT in agriculture holds immense promise. Through these advancements, we can not only enhance the productivity of agriculture but also contribute to a more sustainable and food-secure world for future generations. This transformation in agriculture is a testament to the potential of technology to address critical global challenges.

KEYWORDS

- **smart farming technology**
- **precision agriculture**
- **plant surveillance systems**
- **pest identification solutions**
- **IoT in agriculture**
- **agricultural innovation**
- **advanced farm monitoring**
- **smart pest management**
- **precision crop protection**
- **digital agriculture**

REFERENCES

1. Nehra, V., Sharma, M., & Sharma, V. IoT based smart plant monitoring system. In: *Proceeding of 13th International Conference on Cloud Computing, Data Science & Engineering (Confluence)*, **2023**, pp. 60–65, IEEE.
2. Sambath, M., Prasant, M., BhargavRaghava, N., & Jagadeesh, S. Iot based garden monitoring system. *Proceeding of International Conference on Physics and Photonics Processes in Nano Sciences*, 20–22 June **2019**, Eluru, India.
3. Sahu, C. K., & Behera, P. A low cost smart irrigation control system. In: *Proceeding of 2nd International Conference on Electronics and Communication Systems (ICECS)*, **2015**, pp. 1146–1152. IEEE.
4. Corrales, D. C., Corrales, J. C., & Figueroa-Casas, A. Towards detecting crop diseases and pest by supervised learning. *Ingenieríay Universidad* **2015**, *19*(1), 207–228.
5. Nawaz, M. A. et al. Plant disease detection using Internet of Thing (IoT). *International Journal of Advanced Computer Science and Applications* **2010**, *11*(1).
6. Elemure, I. E., Akinyoola, J. O., Olawumi, A. M., Amoran, A. E., Oshin, E. A., & Adebanwo, T. V. Smart sorting method of grains using a colour sensor. *International Journal of Scientific Research & Engineering Trends* **2019**, *5*(3), 2395–566.
7. Seetharaman, K. A fully automated crop disease monitoring and management system based on IoT: IoT-based disease identification for banana leaf. In: *Deep Learning Applications and Intelligent Decision Making in Engineering,* IGI Global, **2021**, pp. 192–211.
8. Suma, N., Samson, S. R., Saranya, S., Shanmugapriya, G., & Subhashri, R. IOT based smart agriculture monitoring system. *International Journal on Recent and Innovation Trends in Computing and Communication* **2017**, *5*(2), 177–181.

9. Bhadani, P., & Vashisht, V. Soil moisture, temperature and humidity measurement using arduino. *Proceeding of 9th International Conference on Cloud Computing, Data Science & Engineering*, **2019**, pp. 567–571.
10. Priya, C. G., Pandu, M. A., & Chandra, B. Automatic plant monitoring and controlling system over GSM using sensors. In: *Proceeding of IEEE Technological Innovations in ICT for Agriculture and Rural Development (TIAR)*, **2017**, pp. 173–176. IEEE.
11. Priya, C. G., Pandu, M. A., & Chandra, B. Automatic plant monitoring and controlling system over GSM using sensors. In: *Proceeding of IEEE Technological Innovations in ICT for Agriculture and Rural Development (TIAR)*, **2017**, pp. 173–176. IEEE.
12. Suma, N., Samson, S. R., Saranya, S., Shanmugapriya, G., & Subhashri, R. IOT based smart agriculture monitoring system. *International Journal on Recent and Innovation Trends in Computing and Communication* **2017**, *5*(2), 177–181.
13. Thakare, S., & Bhagat, P. H. Arduino-based smart irrigation using sensors and ESP8266 WiFi module. In: *Proceeding of Second International Conference on intelligent Computing and Control Systems (ICICCS)*, **2018**, pp. 1–5. IEEE.
14. Nisha, G., & Megala, J. Wireless sensor Network based automated irrigation and crop field monitoring system. In: *Proceeding of Sixth International Conference on Advanced Computing (IcoAC)* **2014**, pp. 189–194. IEEE.
15. Vaishali, S., Suraj, S., Vignesh, G., Dhivya, S., & Udhayakumar, S. Mobile integrated smart irrigation management and monitoring system using IOT. In: *Proceeding of International Conference on Communication and Signal Processing (ICCSP)*, **2017**, pp. 2164–2167. IEEE.

CHAPTER 9

IoT-Enhanced Energy Storage and Distribution

MANOJ SINGH ADHIKARI[1,*], RAJU PATEL[2], MANOJ SINDHWANI[1], SHIPPU SACHDEVA[1], and MOHIT PAYAL[3]

[1]*School of Electronics and Electrical Engineering, Lovely Professional University, Phagwara, Punjab, India*

[2]*School of Electronics Engineering, Vellore Institute of Technology, Chennai, Tamil Nadu, India*

[3]*Electronics and Communication Engineering, Graphic Era Hill University, Dehradun, Uttarakhand, India*

*Corresponding author

ABSTRACT

A new era of opportunities for the energy sector has begun with the integration of Internet of Things (IoT) technology. The current energy ecosystem is more effective and sustainable results from the integration of IoT. We produce and use energy has transformed dramatically because of smart grids, classy sensors, and predictive analytics. As demand on the energy grid increases, real-time monitoring and control are required. There are many difficulties in IoT-enhanced energy distribution and storage. Reliable cyber security defenses and data privacy policies are crucial. Furthermore, to maintain fair competition, grid reliability, and consumer protection regulatory frameworks must change to reflect this changing environment. This integration improved grid stability, increased energy efficiency, and empowered consumers who actively participate in energy management. A step toward a more sustainable future is being able to

IoT Potential for Green Energy Solutions. Abhishek Kumar, Ajay Roy, & Shamimul Qamar (Eds.)

DOI: 10.1201/9781779643599-9

optimize energy usage and integrate renewable energy sources. Achieving the potential of IoT-enhanced energy systems will require cooperation between researchers and industry leaders. By working together, we can keep modifying, improving, and developing our methods for distributing and storing energy.

9.1 INTRODUCTION

The introduction of the Internet of Things (IoT) into the field of energy distribution and storage represents a revolutionary step toward the creation of a more intelligent, effective, and sustainable energy ecosystem in a time of rapid technological advancement [1–3]. A new era of interconnectedness is being ushered in by the convergence of IoT and energy infrastructure, which makes it possible to monitor, control, and optimize energy resources in real time on an unprecedented scale [4–8]. This combination meets the urgent need for a more flexible and resilient energy system that can both lessen its environmental impact and meet the rising demands of the world's population. We can transform the way we produce, distribute, and use energy by integrating IoT technology into conventional energy storage and distribution systems. This opens a wide range of possibilities [9–12]. This investigation explores the fundamental ideas, available technologies, and possible consequences of IoT-enhanced energy distribution and storage [13, 14]. A patchwork of innovations comes together to form a dynamic and adaptable energy landscape, from smart grids and sophisticated sensors to predictive analytics and blockchain-enabled transactions. Benefits from this landscape are expected to include increased grid stability and dependability as well as empowered consumers who actively engage in energy management [15–18].

We will closely examine the complex interactions that occur between IoT devices, data analytics, and the energy grid's physical infrastructure as we travel. We will also evaluate the crucial factors related to data privacy, cybersecurity, and regulatory frameworks that support the integration of IoT successfully in energy systems [19–22].

As we commence this investigation, it is crucial to acknowledge that the path toward IoT-enhanced energy distribution and storage is a cooperative undertaking, encompassing stakeholders from the research, policy, and industry sectors. By working together, we can create a future in which energy is not only widely available and abundant but also intelligent and sustainable [23–25]. For those who can contribute to this revolutionary journey toward a more intelligent and interconnected energy future, this document acts as a roadmap, a motivator, and a source of inspiration. To increase the efficiency, dependability, and sustainability of energy grid operations, smart technologies and sensors are integrated into IoT-enabled energy storage and distribution concepts [26–30]. The chapter explores key concepts and technological progresses in IoT-enabled energy distribution and storage.

9.2 SMART GRIDS

The IoT is essential to improving energy distribution and storage in smart grids. Smart grids maximize energy production, distribution, and consumption by utilizing cutting-edge technologies. Figure 9.1 shows the integration of IoT in a smart grid system and is discussed as follows:

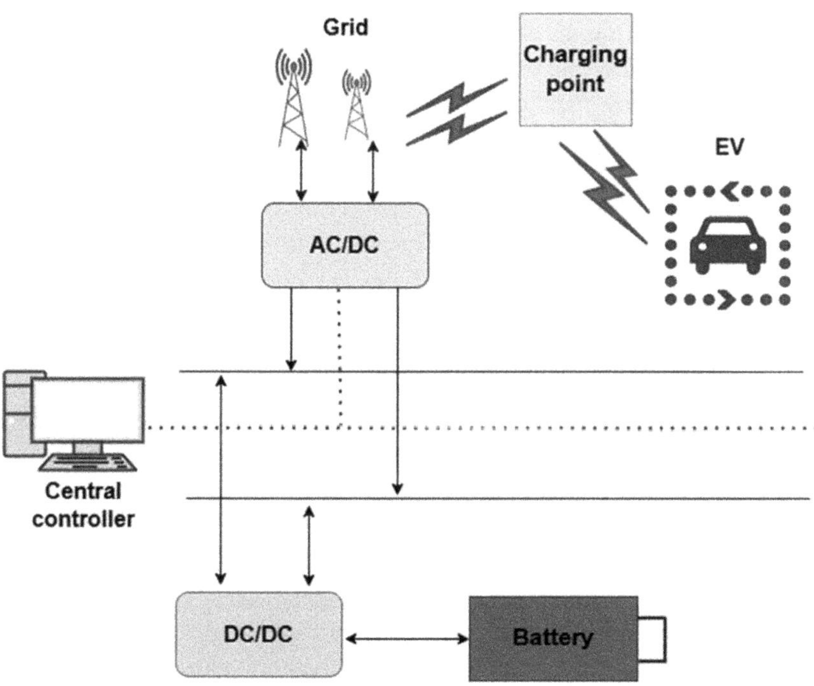

FIGURE 9.1 Integration of IoT in a smart grid system.

- *Real-time monitoring and control*: Energy storage levels, grid efficiency, and equipment conditions can all be tracked in real time by IoT sensors and devices. This information facilitates more effective energy distribution and aids in supply and demand balance.
- *Demand response (DR)*: Smart systems and appliances in homes and businesses can interact with IoT devices. DR programs are made possible by this, allowing energy consumption to be dynamically changed in response to pricing and grid conditions.
- *Predictive maintenance*: Performance data can be gathered by sensors on grid components and energy storage systems (ESS). IoT analytics can predict maintenance needs, helping to reduce downtime and ensure peak performance while also saving costs.

The integration of IoT technology revolutionizes various aspects of the energy sector, contributing to enhanced efficiency and sustainability. First, IoT sensors play a crucial role in optimizing ESS by monitoring battery conditions, temperature, and other relevant parameters. This information enables the system to optimize charging and discharging patterns for maximum efficiency. Second, IoT devices facilitate the integration of renewable energy sources, such as wind turbines and solar panels, by tracking their output. Through the analysis of weather data and grid demand, these gadgets forecast and plan for intermittent energy generation, ensuring seamless integration into the grid. Third, IoT sensors enhance grid resilience by promptly identifying faults or disturbances and notifying users, allowing for swift responses to reduce interruptions and preserve stability. Additionally, IoT devices contribute to grid security by offering features that guard against online attacks, monitoring, and defending vital infrastructure against potential cyber threats. Real-time data on voltage and frequency levels provided by IoT sensors aids in grid balancing and frequency regulation, preserving stability by adjusting generation or storage capacity as needed. Furthermore, IoT data is instrumental in long-term planning by evaluating historical data, forecasting future trends, and identifying areas requiring grid upgrades or expansion. Lastly, IoT-enabled smart grids promote carbon emission reduction by optimizing energy storage and distribution, facilitating the integration of more renewable energy sources and thereby contributing to a greener and more sustainable energy landscape.

Hence, IoT-enhanced smart grid energy distribution and storage are essential for developing more dependable, efficient, and sustainable energy systems. Smart grids can adjust to changing circumstances and guarantee a reliable and consistent supply of energy by utilizing real-time data and sophisticated analytics.

9.3 GRID MANAGEMENT SYSTEMS

Grid management systems are being revolutionized by IoT technology, which offers real-time data and sophisticated control capabilities. Figure 9.2 shows the deep learning for the energy management system. IoT improves energy distribution and storage in grid management systems in the following ways:

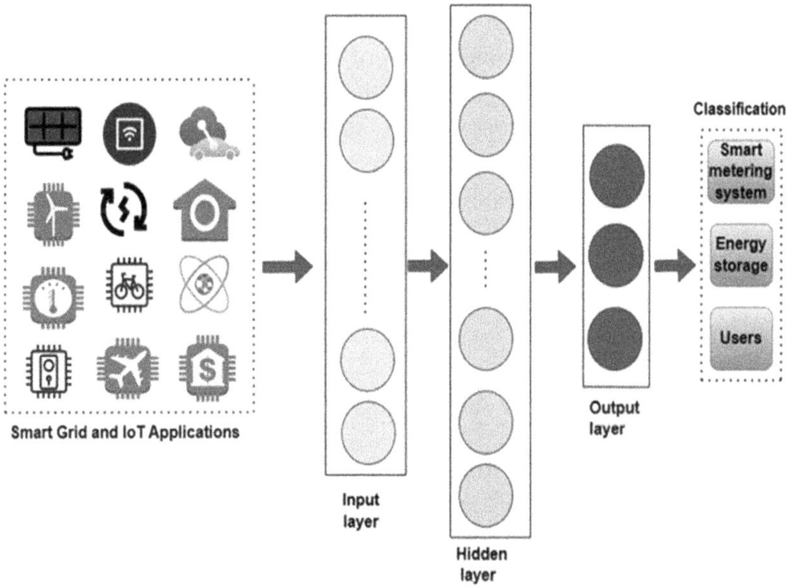

FIGURE 9.2 Deep learning for energy management system.

- *Monitoring and control in real time*: Continuous, real-time data on grid conditions, energy production, and consumption is provided by IoT sensors and devices. This allows grid operators to optimize energy distribution and storage with real-time adjustments.
- *Data analytics and machine learning*: Advanced analytics and machine learning algorithms can be used to process data generated

by the IoT. Grid operators can anticipate future demand, optimize energy distribution, and obtain insights into patterns of energy consumption as a result.

The incorporation of IoT technology into grid management systems brings about several advantages, fostering a more efficient and resilient energy infrastructure. First, IoT-enabled systems excel in DR and load balancing by dynamically shifting loads based on real-time tracking of energy consumption patterns. This capability not only facilitates the implementation of DR programs but also contributes significantly to maintaining grid stability by balancing supply and demand. Second, IoT sensors play a pivotal role in fault detection and rapid response, promptly identifying grid faults or disturbances and triggering immediate alerts. This swift response mechanism enables quick interventions to reduce interruptions and preserve overall grid stability. Third, IoT sensors contribute to the seamless integration of renewable energy sources, such as wind turbines and solar panels, by tracking their output. In situations where renewable energy generation is sporadic, this data is harnessed to forecast and optimize patterns for energy distribution and storage, ensuring efficient utilization of renewable resources. Overall, the integration of IoT in grid management enhances responsiveness, load optimization, and the effective integration of renewable energy, thereby advancing the reliability and sustainability of the energy grid.

9.3.1 GRID FREQUENCY AND VOLTAGE REGULATION

IoT sensors mounted on grid management systems can monitor the frequency and voltage levels of the electrical supply. This information is crucial for preserving grid stability and adjusting generation or storage capacity as needed. Improved Battery and Energy Storage Management Energy storage systems' IoT sensors monitor the health, temperature, and charge level of their batteries. These data are used to optimize charging and discharging strategies for maximum efficiency.

9.3.2 GRID RESILIENCE AND RELIABILITY

The ability of IoT-enhanced grid management systems to react swiftly to disruptions can boost the grid's resilience and dependability.

9.3.3 GRID PLANNING AND EXPANSION

IoT-generated data can be used for long-term planning by analyzing historical data, predicting future trends, and identifying areas that require grid upgrades or expansion.

9.3.4 MARKET INTEGRATION AND ENERGY TRADING

Peer-to-peer energy trading and market participation are possible for IoT-enabled grid management systems. They can independently bargain for energy prices according to supply and demand.

Hence, modernizing energy grids is made possible by IoT-enhanced grid management systems, which offer real-time data and advanced control capabilities. This results in energy distribution and storage that is more sustainable, dependable, and efficient.

9.4 DISTRIBUTED ENERGY RESOURCES

The management and integration of distributed energy resources (DERs) in contemporary energy systems are greatly improved by IoT technologies. A variety of decentralized energy sources, such as microgrids, solar panels, wind turbines, and ESS, are included in DERs. Figure 9.3 shows the IoT-based hybrid AC/DC microgrid system. IoT improves energy distribution and storage in DERs in the following ways:

- *Real-time monitoring and control*: Data on the performance of DERs is available in real time thanks to IoT sensors and devices. This makes it possible for operators to keep an eye on energy production, consumption, and storage levels, enabling quick adjustments to maximize efficiency.
- *Grid integration and stability:* DERs with IoT enhancements can interact with each other and the central grid. Grid stability is aided by this communication, which enables frequency stabilization, voltage control, and grid synchronization.

The integration of IoT technology into DERs brings about a range of benefits that contribute to the efficiency, reliability, and sustainability of energy systems. First, IoT-enabled DERs open avenues for optimized

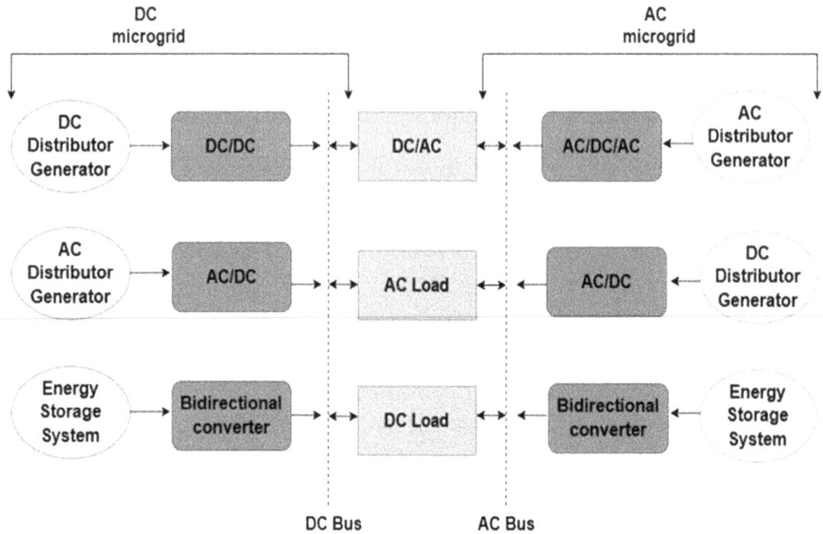

FIGURE 9.3 IoT-based hybrid AC/DC microgrid system.

energy trading, allowing peer-to-peer energy transactions or market participation. These systems autonomously negotiate energy prices based on real-time supply and demand, maximizing profits for both producers and consumers. Second, the implementation of DR and load shifting becomes more effective with IoT capabilities in DERs. These resources can dynamically adjust their energy production or consumption patterns in response to grid signals, facilitating grid balancing during peak demand or supply shortages. IoT sensors provide essential data on weather, energy production, and consumption trends, supporting forecasting and predictive analytics for efficient grid planning. Third, IoT enhances resilience and reliability by continuously monitoring DERs. Quick fault detection and isolation reduce downtime during disruptions, and optimized battery management increases system efficiency and battery life. Fourth, IoT technology ensures data privacy and cybersecurity, protecting sensitive energy data from online threats and unauthorized access. IoT-enabled remote diagnostics and maintenance enhance performance and minimize downtime through proactive maintenance. Lastly, IoT-enabled DERs can operate as stand-alone microgrids or integrate into larger grids, intelligently managing energy generation, consumption, and storage based on local demand and resource availability. This decentralized approach contributes to more effective, reliable, and sustainable energy distribution and storage. Overall, IoT-enhanced DERs play a crucial role in modernizing energy systems by providing real-time data and advanced control capabilities.

9.5 ENERGY STORAGE SYSTEMS

Energy distribution and storage in ESS are improved by IoT technology. ESS includes several technologies, including pumped hydro storage, flywheels, and batteries. Figure 9.4 shows the energy distribution and storage in ESS. IoT improves energy distribution and storage in ESS in the following ways:

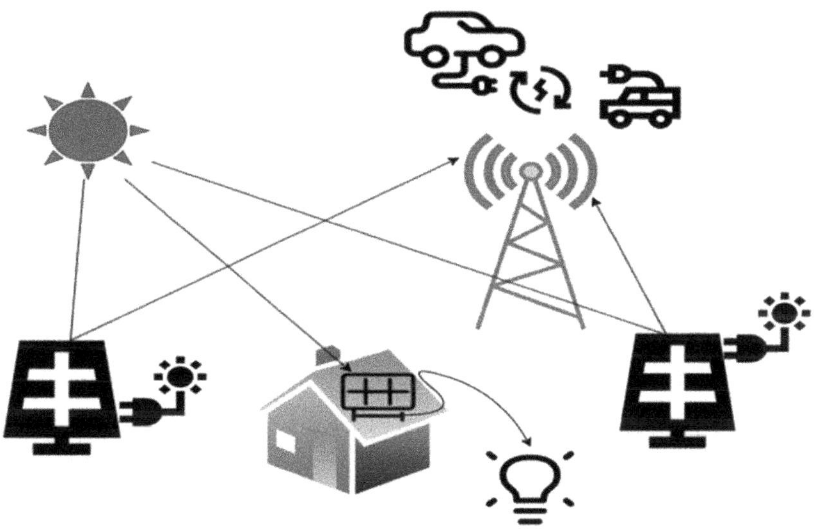

FIGURE 9.4 Energy distribution and storage in ESS.

The incorporation of IoT technology into ESS revolutionizes the monitoring and control of these systems in real time. IoT sensors provide instantaneous data on ESS performance, temperature, and charge level, facilitating precise control and optimization of charging and discharging procedures. Predictive maintenance benefits from IoT-enabled sensors, which gather information on the state and functionality of energy storage components, allowing for anticipatory maintenance and ensuring peak performance while minimizing downtime. Additionally, IoT technology enables seamless integration with electric vehicles and charging stations, facilitating intelligent charging plans to maximize energy efficiency and alleviate grid stress during peak hours. The dynamic adjustment of charging and discharging schedules in response to demand projections, energy prices, and grid conditions is made possible by IoT, leading to more effective use of stored energy. Real-time grid frequency monitoring and stabilization are achieved with IoT sensors on ESS, allowing participation in DR programs. IoT-enabled ESS can modify energy storage and discharge patterns in response to grid signals, contributing to the balance between supply and demand. Furthermore, IoT sensors track the output of renewable energy sources, aiding in the optimization of energy distribution and storage patterns, especially in scenarios of sporadic renewable energy generation. Energy arbitrage becomes feasible as IoT-enabled ESS analyzes market dynamics and grid demand trends, storing energy during periods of low demand and releasing it during high-demand, high-price periods. These systems can function independently as microgrids or collaboratively with the main grid, intelligently managing energy storage based on available resources and local demand.

Overall, IoT-enhanced ESS plays a pivotal role in promoting sustainable energy practices and reducing greenhouse gas emissions through optimized energy distribution and storage. Hence, IoT-enhanced ESS, which offers real-time data and enables advanced control capabilities, is essential to modernizing energy systems. This results in energy distribution and storage that is more sustainable, dependable, and efficient.

9.6 DEMAND RESPONSE

DR programs, which are essential for balancing energy supply and demand on the grid, are greatly enhanced by IoT technology. Figure 9.5 shows the IoT monitoring DR system. IoT improves energy distribution and storage in DR in the following ways:

The integration of IoT technology into DR programs brings about a paradigm shift in how energy distribution and storage are managed during grid events and peak demand periods. IoT sensors play a crucial role in providing real-time data on equipment performance, grid conditions, and energy consumption, forming the foundation for prompt reactions to grid events, a cornerstone of DR programs. Devices with IoT capabilities can autonomously respond to DR signals from grid operators, adjusting

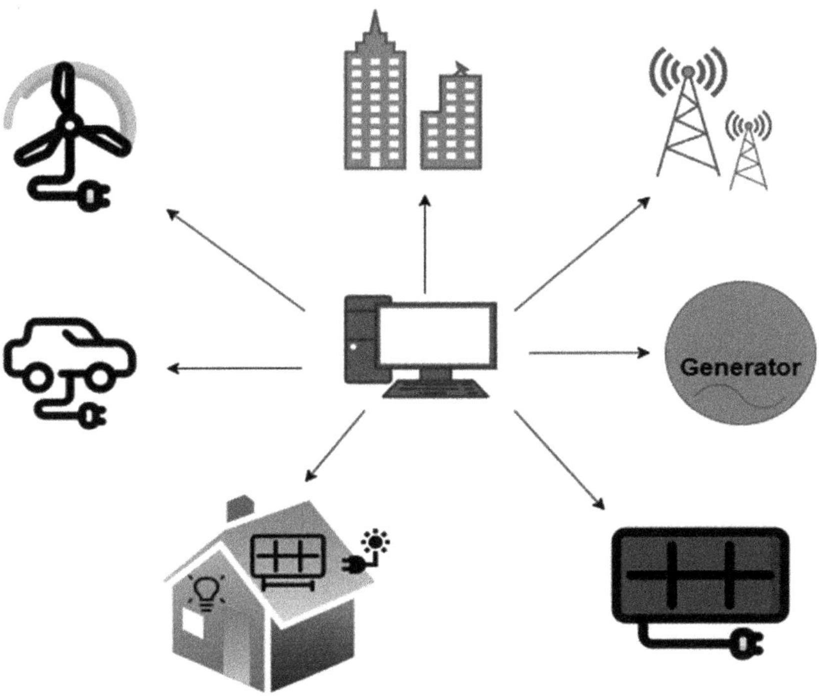

FIGURE 9.5 IoT monitoring DR system.

nonessential loads, modifying energy usage, or activating energy storage devices as needed. The direct communication between IoT-enabled smart meters and DR platforms enhances response accuracy and provides precise energy usage data. ESS and IoT-enabled devices dynamically alter their operations, facilitating load shifting and peak shaving to alleviate strain on the grid during periods of high demand. Additionally, IoT sensors on energy storage devices monitor demand projections, grid conditions, and charge levels, optimizing charging and discharging in alignment with DR regulations. The tracking of renewable energy sources, such as wind and solar power, by IoT devices ensures maximized production during DR events, contributing to grid stability. Remote control and management capabilities enable the dynamic adjustment of energy storage and consumption in response to DR events. Real-time information delivered to customers through IoT-enabled interfaces fosters engagement, promoting energy efficiency and conservation. IoT sensors further contribute to grid stability by monitoring conditions and triggering disaster recovery events during emergencies. Participants in IoT-enhanced distributed generation programs may benefit from financial incentives, encouraging economic advantages. Data analytics and reporting, facilitated by IoT-generated data, provide insights for program improvement, resulting in a more effective and enhanced DR program. In conclusion, IoT-enhanced DR programs leverage real-time data and advanced control capabilities to effectively manage energy distribution and storage, enhancing the robustness, reliability, and efficiency of the energy system.

9.7 INTEGRATION WITH RENEWABLE ENERGY SOURCES

Modernizing energy systems must include integration with renewable energy sources. This process entails integrating eco-friendly, sustainable energy production techniques into already-existing grids. Figure 9.6 shows the establishment of a green energy system. The process of integrating renewable energy sources is as follows:

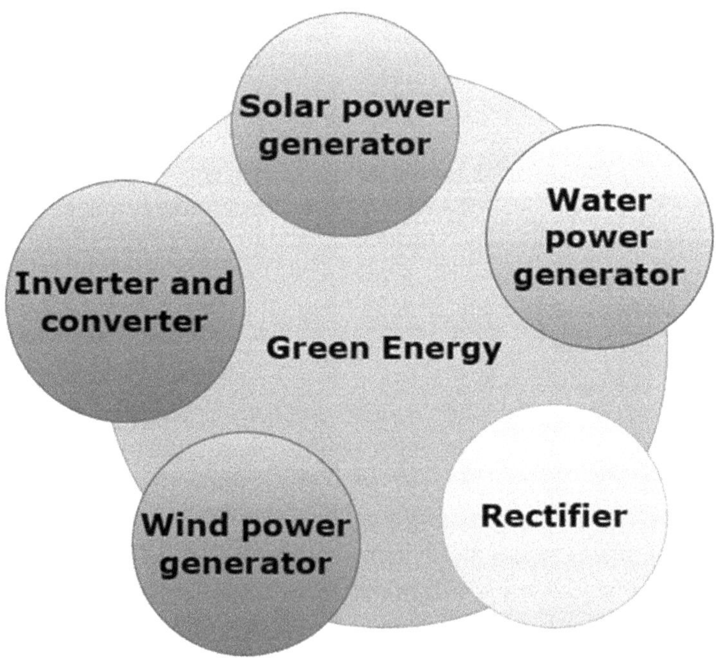

FIGURE 9.6 Establishment of green energy.

- *Wind turbines:* By utilizing wind energy, wind turbines produce electricity. They can be put in place offshore or onshore in windy environments. Usually, the generated electricity is added to the grid.
- *Photovoltaic systems in the Sun:* Sunlight is converted into electricity by solar panels. They can be incorporated into building materials or put on solar farms or rooftops. The electricity produced can be used locally or fed back into the grid.
- *Hydroelectric power:* Hydroelectric power plants produce electricity by utilizing the flow of water. Tidal and wave energy, run-of-river systems, and dams can all be used to accomplish this. The electricity produced is incorporated into the grid.
- *Geothermal energy:* To produce electricity, geothermal power plants harness the natural heat of the Earth. Usually, this is carried out in areas where geothermal activity is active. The energy generated is incorporated into the grid.
- *Biomass and bioenergy*: Biomass energy comes from waste products and organic materials such as wood. It can be directly used to generate electricity and integrated into the grid, or it can be processed into biogas or biofuels.
- *Frequency regulation and grid balancing*: Renewable energy sources, especially solar and wind, can have sporadic output. Grid management systems balance supply and demand and control frequency by utilizing technologies such as energy storage, DR, and grid-scale batteries.
- *Sophisticated control systems and advanced grid management*: To control the unpredictability of renewable energy sources, grid operators use sophisticated control systems. To optimize grid operations, these systems make use of forecasting, real-time data, and predictive analytics.
- *Grid-interactive technologies and smart inverters:* In solar PV systems, smart inverters regulate the electricity flow between the solar panels and the grid. For improved integration, they can modify voltage levels and react to grid conditions.
- *Microgrids with renewable energy:* Microgrids are localized energy systems that frequently incorporate renewable energy sources to give certain towns, campuses, or industrial facilities a dependable and sustainable energy supply.
- *Grid codes and standards:* To guarantee the secure and dependable integration of renewable energy sources into the current

grid infrastructure, grid codes and industry standards have been established. *Renewable energy certificates (RECs) and feed-in tariffs:* To encourage the integration of renewable energy sources, governments frequently enact policies like feed-in tariffs or RECs, which offer financial incentives or guarantees for the production of renewable energy.

Hence, the process of incorporating renewable energy sources into current energy systems is complex and calls for careful planning, cutting-edge grid management tools, and supportive legislation. We can lower greenhouse gas emissions, improve energy sustainability, and build a more robust and effective energy infrastructure by utilizing renewable resources.

9.8 SMART METERS

Smart meters play a crucial role in the IoT ecosystem, enhancing energy storage and distribution in the modern energy landscape. The benefits of IoT-enabled smart meters are numerous, contributing significantly to efficient energy management. IoT technologies improve energy distribution and storage in smart meters in the following ways:

Smart meters equipped with IoT capabilities play a pivotal role in revolutionizing the management and efficiency of energy systems. Offering real-time data monitoring, these smart meters provide up-to-date information on energy output and consumption, enabling grid managers to balance supply and demand effectively. The remote monitoring and control capabilities facilitated by IoT technology empower utilities to conduct operations seamlessly, from firmware updates to configuration modifications and remote disconnects/reconnects. IoT-enabled smart meters also facilitate DR and load-shifting programs by interacting with household and commercial appliances, allowing dynamic adjustments in response to pricing and grid conditions. Moreover, they contribute to grid health monitoring by tracking power supply quality and identifying irregularities or disruptions, ensuring grid dependability. Integration with renewable energy sources enables the tracking of solar panels and other renewables, optimizing energy generation forecasts and grid operations. Additionally, IoT-enhanced smart meters utilize advanced analytics for energy theft detection, swiftly identifying and resolving anomalies in consumption patterns. Voltage and frequency regulation capabilities contribute to fine-tuning energy distribution for grid stability. Security features in IoT technology ensure cybersecurity and data privacy, safeguarding smart meter data against cyber threats. By providing consumers with real-time energy usage data, smart meters enabled by IoT technology empower individuals to make well-informed choices regarding energy efficiency and conservation. Lastly, by maximizing energy usage and integrating renewable sources, IoT-enhanced smart meters contribute to the reduction of carbon emissions, promoting sustainable energy practices.

Hence, by offering real-time data and enabling sophisticated control capabilities, IoT-enhanced smart meters are essential to modernizing energy management. This results in smart grid energy distribution and storage that is more sustainable, dependable, and efficient.

9.9 PREDICTIVE ANALYTICS

Predictive analytics and IoT technology work well together to optimize energy distribution and storage. Using historical data, current information, and machine learning algorithms, predictive analytics helps users predict future trends and make well-informed decisions. When implemented in energy systems, it can result in distribution and storage that is more dependable and efficient. IoT devices play a crucial role in revolutionizing the energy landscape through comprehensive data collection and integration. These devices gather extensive data on energy production, consumption, grid conditions, and equipment performance, forming the foundation for predictive analytics models. Real-time data feeds from IoT enable swift adjustments to energy distribution and storage plans based on changing circumstances. Predictive analytics, fueled by IoT data, facilitates demand forecasting, allowing grid operators to efficiently allocate energy resources and plan for peak usage periods. IoT sensors also track renewable energy sources' output, supporting predictive analytics in optimizing energy storage and distribution strategies. Meteorological data collected by IoT aids in weather-based predictions, enhancing planning for renewable energy generation and consumption patterns. Furthermore, IoT-enabled sensors monitor the state of grid and energy storage equipment, enabling predictive analytics to schedule maintenance efficiently and ensure peak performance. Load forecasting, battery health management, risk assessment, and contingency planning are all areas where IoT data and predictive analytics converge, providing valuable insights for effective energy system management. Additionally, the combination of predictive analytics and IoT data optimizes energy trading strategies, empowering traders to make informed decisions for maximizing profits. Ultimately, predictive analytics contribute to carbon emission reduction by integrating more renewable energy sources and aligning energy storage and distribution with environmental sustainability goals.

Hence, IoT and predictive analytics work together to improve energy systems' capabilities by supplying precise, up-to-date data that helps with decision-making. This results in energy storage and distribution methods that are more effective, dependable, and sustainable.

9.10 GRID-EDGE DEVICES

Enhancing energy storage and distribution at the grid edge—the outer edge of the electrical grid where consumers' energy resources and loads interact with the larger grid—is made possible in large part by IoT technology. IoT improves energy distribution and storage in grid-edge devices in the following ways: grid-edge IoT devices contribute significantly to real-time monitoring and control of energy systems. These devices, equipped with sensors, provide instantaneous information on energy generation, consumption, and grid conditions, enabling precise regulation and optimization of energy use. Furthermore, grid-edge IoT-enabled devices actively participate in DR initiatives, automatically adjusting their energy consumption in response to changes in demand and grid signals. These devices play a crucial role in ensuring grid stability and support by monitoring parameters such as voltage and frequency, independently delivering grid support functions like frequency stabilization and voltage control. Additionally, grid-edge devices with energy storage capabilities leverage IoT technology to optimize battery charging and discharging based on demand patterns and grid conditions. The integration of machine learning algorithms into IoT devices at the grid edge enhances their intelligence, allowing them to assess local data and make informed decisions about energy storage and use. This intelligence extends to optimal energy trading, where grid-edge IoT devices can engage in peer-to-peer energy trading or market participation, independently bargaining for energy prices based on supply and demand. Overall, grid-edge IoT devices play a pivotal role in enhancing the efficiency, responsiveness, and intelligence of energy systems.

Hence, IoT-enhanced grid-edge devices, which offer real-time data and enable advanced control capabilities at the edge of the grid, are vital to the modernization of energy systems. This results in energy distribution and storage that is more sustainable, dependable, and efficient.

9.11 CYBERSECURITY AND DATA PRIVACY

In the context of energy storage and distribution, IoT technology presents both new cybersecurity and data privacy opportunities as well as challenges. To guarantee that data transferred between devices and systems is safe and cannot be intercepted or tampered with, IoT devices use encrypted communication protocols. Strong authentication and authorization processes are used by IoT devices to guarantee that only systems or individuals with permission can access sensitive information and perform control operations. Every IoT device is given a distinct identity and cryptographic key, which makes it possible for the network to validate and trust it. This lessens the possibility of data manipulation or access by unauthorized devices. Firmware updates and secure boot procedures are used by IoT devices to guarantee that only verified and unmodified firmware can be loaded. To further guard against tampering, firmware updates are digitally signed and encrypted. IoT systems employ access control mechanisms to limit which users or systems have access to data and functions. These mechanisms include role-based permissions and access control. Permissions based on roles guarantee that users possess only the requisite rights for their roles. To keep an eye out for questionable activity or attempts to compromise security, IoT systems are outfitted with intrusion detection and prevention systems. This may entail spotting odd trends in data traffic or login attempts. Even if physical access to an IoT device is obtained, confidential data is encrypted and stored on the device to prevent unwanted access. Network segmentation and fire walls are used by IoT networks to separate distinct network segments and stop illegal access to vital systems and data.

- *Continuous auditing and monitoring:* IoT systems keep an eye out for unusual activity and security incidents. For auditing purposes, logs are created and kept, which makes it possible to look into security incidents.
- *Planning for incident response and recovery:* IoT systems are equipped with plans for quickly responding to security incidents.

This covers containment, eradication, recovery, and takeaways for preventive measures in the future.

IoT systems are made to abide by industry-specific cybersecurity and data privacy standards and regulations, guaranteeing legal and regulatory compliance.

To prevent inadvertent security breaches, all stakeholders, including individuals and operators who are well-versed in cybersecurity and data privacy protocols, must be adequately trained. Hence, cybersecurity and data privacy are given top priority in IoT-enhanced energy storage and distribution systems. This is achieved through a combination of technological measures, secure protocols, access controls, and ongoing monitoring. This promotes a safe and dependable energy ecosystem by guaranteeing that vital energy infrastructure is shielded from online attacks and unauthorized access.

9.12 BLOCKCHAIN FOR ENERGY TRANSACTIONS

In the context of energy storage and distribution, blockchain technology in conjunction with IoT technology provides a potent solution for safeguarding and streamlining energy transactions. Blockchain technology offers a decentralized ledger that documents every energy-related transaction. Since this ledger is transparent, all parties involved can see and check transactions. The integrity of transaction history is guaranteed by the blockchain's inability to change data once it has been recorded. Energy transactions can be automatically triggered and verified by smart contracts running on a blockchain, provided certain conditions are met. Smart contracts can communicate with gadgets and sensors in conjunction with the IoT to make sure all requirements are satisfied before completing a transaction. Real-time data on energy production, consumption, and grid conditions can be gathered by IoT sensors. Peer-to-peer energy trading on a blockchain is made possible by this data, enabling customers to buy and sell excess energy directly. Data regarding the energy generation source (such as fossil fuels or renewable energy sources) can be obtained through IoT sensors. Customers will be able to select energy according to its source if this data is stored on the blockchain. Local energy production, consumption, and storage can be observed by IoT devices within a microgrid. The microgrid's energy can be automatically distributed by smart contracts on the blockchain following preset guidelines.

Data on demand and conditions on the grid can be obtained from IoT sensors. Grid service markets that are based on blockchain technology can make use of this data. Smart contracts and IoT data can automate the processes of invoicing and settlement. This lowers the possibility of disputes and does away with the need for middlemen. An extra degree of protection can be added by encrypting and storing IoT-generated data on a blockchain. Private keys provide access control over sensitive energy data. Real-time monitoring of the grid's voltage and frequency levels is possible thanks to IoT sensors. Blockchain technology combined with this data can be used to automatically modify energy production and consumption to preserve grid stability. The carbon emissions resulting from energy production can be monitored by IoT sensors. To validate emissions reductions and support carbon credit programs, this data can be stored on a blockchain.

Blockchain technology offers a transparent and auditable record of energy transactions, facilitating regulatory compliance and auditing. This can help auditing procedures and streamline efforts to comply with regulations. Hence, energy transactions are revolutionized when IoT and blockchain technology are combined because they offer automation, security, and transparency. In the current energy landscape, this allows for more sustainable, dependable, and efficient energy distribution and storage.

9.13 ENERGY EFFICIENCY AND CONSERVATION

Two essential elements of sustainable energy practices are conservation and energy efficiency. To achieve positive effects on the economy, environment, and society, they entail making effective use of energy resources and reducing waste. Purchasing appliances and equipment with energy labels or ratings that are higher than the Energy Star program can help save energy usage. Compared to conventional incandescent bulbs, LED bulbs provide greater energy efficiency and longer operational lives. Properly insulated and sealed structures can reduce heating and cooling requirements by retaining warmth during the winter and keeping spaces cool in the summer. They can be added to enhance insulation and decrease energy wastage. These windows can have double or triple glazing and low-emissivity coatings. By modifying temperatures in response to occupancy patterns and outside meteorological conditions, smart thermostats maximize heating and cooling. To maintain comfortable indoor temperatures, high-efficiency heating, ventilation, and air conditioning (HVAC) systems require less energy. Cogeneration and Combined Heat and Power systems increase overall energy efficiency by simultaneously producing useful heat and electricity from a single energy source.

To optimize production processes and cut down on energy usage, industries can adopt energy-efficient practices and technologies. Using public transportation, carpooling, and switching to electric or hybrid vehicles can all help cut down on emissions and fuel use. District energy systems increase overall energy efficiency by using a central plant to provide power, heat, and/or cooling to multiple buildings. Recovering and repurposing waste heat from power plants or industrial operations enhances energy efficiency all around. Energy requirements can be greatly decreased by designing buildings with features like natural ventilation, passive solar design, and effective layouts.

It is said that energy conservation is ensuring that doors and windows are properly sealed and that walls and roofs are well-insulated can help minimize energy loss. This is known as proper weatherization. Regular maintenance guarantees that automobiles, HVAC systems, and appliances run smoothly and use less energy. Phantom energy usage can be avoided by using power strips or unplugging devices while not in use. Encouraging responsible energy use is achieved through educating individuals, households, and businesses about energy-saving practices. Energy audits help prioritize energy-saving projects and pinpoint areas that need improvement. Encouraging responsible energy use is achieved by taking part in energy conservation programs, incentives, and initiatives offered by governments and industries.

Including renewable energy sources: Reducing dependency on fossil fuels can be achieved by utilizing on-site renewable energy generation, such as solar panels.

Encouraging walking, cycling, or taking public transportation can help promote sustainable transportation by lowering the amount of energy used when driving a personal vehicle.

Using water-saving techniques and technologies lowers the energy needed to heat and distribute the water. Hence, people, homes, companies, and industries can all help create a more resilient and sustainable energy future while lowering costs and having a smaller negative impact on the environment by putting energy efficiency and conservation measures into practice.

9.14 CONCLUSION

An exciting new era of opportunities for the energy sector has begun with the integration of IoT technology into energy storage and distribution systems. An energy ecosystem that is more intelligent, effective, and sustainable could result from this convergence. How we produce, handle, and use energy has changed dramatically because of the introduction of smart grids, sophisticated sensors, and predictive analytics. With the increasing demands on the energy grid, real-time monitoring and control have become not only possible but also crucial. There are obstacles in the way of this IoT-enhanced energy distribution and storage, though. Strong cybersecurity defenses and impenetrable data privacy policies are essential. Furthermore, to maintain fair competition, consumer protection, and grid reliability, regulatory frameworks must change to reflect this changing environment.

This integration has many potential advantages. Among the many benefits are increased energy efficiency, improved grid stability, and empowered consumers who actively participate in energy management. A big step toward a more sustainable future is being able to optimize energy usage and seamlessly integrate renewable energy sources. Looking ahead, achieving the full potential of IoT-enhanced energy systems will require cooperation between researchers, legislators, and industry leaders. By working together, we have the opportunity to continuously refine and enhance our energy distribution and storage techniques, leading toward a future where energy is not only affordable and easily accessible but also intelligent and environmentally sustainable.

KEYWORDS

- **IoT**
- **renewable energy**
- **sustainable energy**
- **solar energy**

REFERENCES

1. Lin, C. C., Deng, D. J., Kuo, C. C., & Liang, Y. L. Optimal charging control of energy storage and electric vehicle of an individual on the internet of energy with energy trading. *IEEE Transactions on Industrial Informatics* **2018**, *14*(6), 2570–2578.
2. Liu, D., Chen, Y., Chai, K. K., Zhang, T., & Elkashlan, M. Two-dimensional optimization on user association and green energy Allocation for HetNets with Hybrid Energy Sources. *IEEE Transactions on Communications* **2015**, *63*(11), 4111–4124.
3. Tahsin, A., Roy, P., Razzaque, M. A., Rashid, M. M., Siraj, M., AlQahtani, S. A., Hassan, M. R., & Hassan, M. M. Energy cooperation among sustainable base stations in multi-operator cellular networks. *IEEE Access* **2023**, *11*, 19405–19417.
4. Jahid, A., Hossain, M. S., Monju, M. K. H., Rahman, M. F., & Hossain, M. F. Techno-economic and energy efficiency analysis of optimal power supply solutions for green cellular base stations. *IEEE Access* **2020**, *8*, 43776–43795.
5. Sadiq, M., et al. Future greener seaports: A review of new infrastructure, challenges, and energy efficiency measures. *IEEE Access* **2021**, *9*, 75568–75587.
6. Chen, C., He, B., & Tang, X. Green-aware work load scheduling in geographically distributed data centers. In: *Proceedings of IEEE International Conference on Cloud Computing Technology and Science*, **2012**, pp. 82–89.
7. Adhikari, M. S., Khan, P. S., Senapathi, G., Chatterjee, B., Reddy, D. V., & Malik, P. K. Design of an IOT based smart medicine bbox. In: *Proceedings of IEEE International Conference on Devices for Integrated Circuit (DevIC)*, **2023**, pp. 346–349.
8. Homsi, S., Liu, S., Chaparro-Baquero, G. A., Bai, O., Ren, S., & Quan, G. Workload consolidation for cloud data centers withguaranteed qos using request reneging. *Proceedings of IEEE Transactions on Parallel and Distributed Systems* **2016**, *28*(7), 2103–2116.
9. Sekhar, J., Jeba, G., & Durga, S. A survey on energy efficient serverconsolidation through vmlive migration. *International Journal of Advances in Engineering & Technology* **2012**, *5*(1), 515.
10. Adhikari, M. S., Ramalinga, K. M., Anand, A., Venkatesan, G. P., Ravi, A. R., & Thakur, N. IOT based precision Agri-Bo. In: *IEEE International Conference on Devices for Integrated Circuit (DevIC)*, **2023**, pp. 321–324.
11. Joshi, V., & Adhikari, M. S. IOT-based technology for smart farming. *Electronic Devices and Circuit Design: Challenges and Applications in the Internet of Things*. CRC Press, **2022**, pp. 223–242.
12. Joshi, V., & Adhikari, M. S. Advanced agricultural-based IoT technology. *Intelligent Communication and Automation Systems*. CRC Press, **2021**, pp. 239–254.
13. Billah, S. M. B., Kabir, K. M., Islam, M. O., Barua, S., Mahmud, M. S., & Hossain, M. S. Hydrogen energy storage based green power plant in seashore of Bangladesh: Design and optimal cost analysis. In: *International Conference on Innovations in Green Energy and Healthcare Technologies (IGEHT)*, **2017**, pp. 1–5.
14. Bhattacharya T., & Qin, X. Modeling energy efficiency of future green data centers. *International Green and Sustainable Computing Workshops (IGSC)*, **2020**, pp. 1–3.
15. Delinchant, B., Wurtz, F., Ploix, S., Schanen J. L., & Marechal, Y. Green-ER living lab: A green building with energy aware occupants. In: *International Conference on Smart Cities and Green ICT Systems (SMARTGREENS)*, **2016**, pp. 1–8.

16. Dos Reis, T. N. F., Teixeira, M. M., & Salles Soares Neto, C. D. Relation of energy consumption in green cloud computing with big data. In: *IEEE Green Technologies Conference,* Denver, CO, USA, **2023**, pp. 280–283.
17. Kumar, N. A study on green energy powered cognitive radio network for communication network architecture of smart grid. *International Innovative Applications of Computational Intelligence on Power, Energy and Controls with their Impact on Humanity (CIPECH)* **2018**, *3–7*, 2018.
18. Ma, X., Huang, S., Li, Z., Yang, Z., Li, F., & Chen, L. Quantitative evaluation method of green power development level. In: *International Conference on Power and Energy Systems (ICPES),* **2020**, pp. 420–425.
19. Sithan, M., & Lai, L. L. Application of green technologies in developing countries — reduced carbon emission and conservation of energy. *IEEE Power and Energy Society General Meeting,* **2011**, pp. 1–7.
20. Yarlagadda, N., Verma, Y. K., Adhikari, M. S., & Mishra, V. Opto-VLSI devices and circuits for biomedical and healthcare applications. In: *A Review of Multi-material, Multi-gate MOSFET Structures,* CRC Press, **2023**, pp. 29–40.
21. Ma, X., Huang, S., Li, Z., Yang, Z., Li, F., & Chen, L. Quantitative evaluation method of green power development level. In: *International Conference on Power and Energy Systems (ICPES),* **2020**, pp. 420–425.
22. Entezari, A., Aslani, A., Zahedi, R., & Noorollahi, Y. Artificial intelligence and machine learning in energy systems: A bibliographic perspective. *Energy Strategy Reviews* **2023,** *45*, 101017.
23. Legala, A., & Li, X. Hybrid data-based modeling for the prediction and diagnostics of Li-ion battery thermal behaviours. *Energy and AI* **2022,** *10*, 100194.
24. Machlev, R., Heistrene, L., Perl, M., Levy, K. Y., Belikov, J., Mannor, S., & Levron, Y. Explainable Artificial Intelligence (XAI) techniques for energy and powersystems: Review, challenges and opportunities. *Energy and AI* **2022**, *9*, 100169.
25. Kharlamova, N., Hashemi, S., & Træholt, C. Data-driven approaches for cyber defence of battery energy storage systems. *Energy and AI* **2021**, *5*, 100095.
26. Fan, Q., Ansari, N., & Sun, X. Energy driven avatar migration in green cloudlet networks. *IEEE Communications Letters* **2017**, *21*(7), 1601–1604.
27. Dayarathna, M., Wen, Y., & Fan, R. Data center energy consumption modeling: A survey. *IEEE Communications Surveys Tutorials* **2016**, *18*(1), 732–794.
28. Goyal, Y., Arya, M. S., & Nagpal, S. Energy efficient hybrid policy in green cloud computing. In: *International Conference on Green Computing and Internet of Things (ICGCIoT)*, **2015**, pp. 1065–1069.
29. Tripathi, S. L., Ganguli, S., Kumar, A., & Magradze, T. *Intelligent Green Technologies for Sustainable Smart Cities*. Wiley, **2022**.
30. Umapathy, P., Hemanth, J., Khera, S., Inbamani, A., & Tripathi, S. L. *Machine Learning and the Internet of Things in Solar Power Generation.* CRC Press, **2023**.

CHAPTER 10

IoT and Its Impact on Transportation

DHRUV B. TRIVEDI[1,*], VAISHALI N. MAHAVAR[2], and
NAMAN N. BHAVSAR[3]

[1]*Civil Engineering Department, Rai University, Ahmedabad, Gujarat, India*

[2]*Aartav Engineering Private Limited, Gandhinagar, Gujrat, India*

[3]*Civil Engineering Department, GTU-ITR, Mehsana, Gujarat, India*

*Corresponding author

ABSTRACT

The Internet of Things (IoT) has emerged as a disruptive force, permeating various sectors with its transformative potential. In the realm of transportation, IoT presents an unprecedented opportunity to revolutionize how we conceive, manage, and interact with our transportation systems. In this context, the transportation industry faces multifaceted challenges. Traffic congestion plagues urban centers, jeopardizing productivity and air quality. Road safety remains a persistent concern, with thousands of accidents and fatalities occurring annually. Simultaneously, the environmental repercussions of transportation, notably greenhouse gas emissions, are a growing global concern. This chapter embarks on a comprehensive exploration of IoT's profound impact on transportation, addressing these challenges head-on. We delve into the core elements of IoT, laying the foundation for a deep dive into its transportation applications. IoT transcends mere technology; it epitomizes the era of "smart transportation." By leveraging sensors, real-time data communication, and sophisticated analytics, IoT endows transportation systems with intelligence, efficiency, and sustainability. IoT's impact extends to the realm of sustainability, transforming transportation into a greener, more eco-conscious enterprise.

IoT Potential for Green Energy Solutions. Abhishek Kumar, Ajay Roy, & Shamimul Qamar (Eds.)

DOI: 10.1201/9781779643599-10

The chapter shines a spotlight on its role in route optimization, emissions reduction, and the promotion of eco-friendly transportation alternatives.

10.1 INTRODUCTION TO IoT IN TRANSPORTATION

The dawn of the 21st century has witnessed the emergence of an extraordinary technological revolution known as the Internet of Things (IoT). The concept of connecting everyday objects to the Internet, thereby enabling them to send and receive data, has unleashed a wave of innovation across various sectors. One domain where IoT is making a transformative impact is transportation.

10.1.1 DEFINITION AND CONCEPT OF IoT

At its core, the IoT is an intricate web of interconnected devices, sensors, and systems, all sharing data and functioning in unison to deliver valuable insights and automation. These devices can range from smartphones and wearables to sophisticated sensors embedded in vehicles, traffic lights, and infrastructure. IoT operates on a simple premise *If it can be connected, it can be controlled, monitored, and optimized.*

In the context of transportation, IoT introduces a new paradigm in which vehicles, roads, and various aspects of the transportation network are imbued with digital intelligence. This means that cars, buses, trains, and even bicycles can be equipped with sensors that collect and transmit data on their performance, location, and status. Simultaneously, traffic management systems and infrastructure such as traffic lights, signage, and roadways can become smart, enabling real-time monitoring and control [1].

The concept of IoT in transportation goes beyond connectivity; it embodies a holistic approach to redefining mobility. This transformation allows for more efficient resource allocation, enhanced safety, and greater sustainability.

10.1.2 RELEVANCE AND SIGNIFICANCE IN TRANSPORTATION

IoT's relevance and significance in transportation are hard to overstate. The modern transportation landscape is characterized by a myriad of challenges, ranging from traffic congestion and road safety concerns to environmental sustainability. These issues have long-posed dilemmas for policymakers, transportation agencies, and commuters alike.

One of the most pressing problems is traffic congestion. As urban populations grow, city roads often become clogged with vehicles, leading to lost productivity, wasted time, and increased air pollution. IoT, with its real-time data collection and analysis capabilities, offers a solution to this age-old problem. By equipping roads, vehicles, and infrastructure with IoT sensors, cities can monitor traffic patterns, identify congestion hotspots, and adjust traffic flow in real time [2–5]. This not only reduces travel times but also contributes to a reduction in greenhouse gas emissions.

Road safety is another paramount concern in transportation. Thousands of accidents and fatalities occur annually, and this human cost is often linked to factors like distracted driving, impaired visibility, and reckless behavior. IoT technology plays a pivotal role in enhancing vehicle safety. Vehicles can communicate with each other through Vehicle-to-Vehicle (V2V) technology, sharing data on speed, location, and direction to enhance collision prevention. Moreover, collision avoidance systems utilize the IoT to identify potential dangers and activate emergency responses. In this chapter, we will delve into the fundamentals of IoT, explore its architecture in transportation, and highlight the evolution of transportation into the era of "smart transportation." We will also discuss its critical role in traffic management, vehicle safety enhancement, and sustainability in transportation. Furthermore, we intend to tackle the obstacles and considerations linked with the implementation of IoT, encompassing issues related to data security and privacy. Finally, we will explore the future trends and emerging technologies that promise to redefine the transportation landscape [6].

10.2 IoT FUNDAMENTALS

The IoT is a technological revolution that has taken the world by storm, and it is not limited to any single domain. Its ability to connect devices and enable data exchange has the potential to transform transportation into a smarter, more efficient system. In this chapter, we will explore the key components of IoT and the architecture-specific to its applications in the realm of transportation.

10.2.1 KEY COMPONENTS OF IoT

At the heart of IoT are several key components that work together to enable the seamless flow of data and information. Understanding these components is vital to comprehending how IoT functions and how it can be harnessed for various applications. In the context of IoT in transportation, these components come together in a way that revolutionizes the industry [7–10]. Here are the essential elements:

10.2.1.1 SENSORS AND ACTUATORS

Sensors are the eyes and ears of the IoT system. They detect changes in the physical environment and convert them into digital data. In transportation, these sensors can be found in vehicles, on roads, and in infrastructure. For example, vehicle sensors can monitor speed, location, fuel consumption, and more. Roadside sensors can measure traffic flow, temperature, and weather conditions. Actuators, on the other hand, enable the system to take action based on the data received. For instance, traffic lights can change their timing in response to traffic data, thus optimizing traffic flow.

10.2.1.2 USER INTERFACE

A critical aspect of IoT is the user interface, which allows people to interact with the system. In transportation, this can be as simple as a smartphone app that provides real-time traffic information to drivers or more complex systems used by traffic management authorities to control signals and monitor traffic conditions.

10.2.2 IoT ARCHITECTURE IN TRANSPORTATION

The architecture of IoT in transportation is a structured framework that defines how the components we discussed earlier work together to create a functional system. It is crucial to have a clear understanding of this architecture to appreciate how IoT can address the specific challenges in the transportation sector.

10.2.2.1 EDGE DEVICES

In transportation, the edge devices are the sensors and actuators located on or in vehicles, infrastructure, and roads. They collect data from the physical environment and transmit it to the central system for processing.

10.2.2.2 CONNECTIVITY LAYER

The connectivity layer involves how data is transmitted between the edge devices and the central system. In transportation, this layer utilizes cellular networks, Wi-Fi, and other communication technologies to ensure seamless data exchange.

10.2.2.3 DATA PROCESSING AND ANALYSIS LAYER

This layer is tasked with gathering and processing data received from edge devices. In the context of transportation, it could entail cloud-based servers analyzing real-time information on traffic patterns, weather conditions, and vehicle data to inform decision-making.

10.2.2.4 APPLICATION LAYER

The application layer is where the data and analysis are put to practical use. In transportation, this layer includes traffic management systems, navigation apps, safety features, and more, all aimed at improving the efficiency and safety of transportation.

Comprehending the architecture [11] of IoT in transportation is crucial for a thorough understanding of the seamless integration of data collection, analysis, and real-time decision-making within the realm of smart transportation systems. In the subsequent sections of this chapter, we will explore how these components and architecture come together to revolutionize transportation, addressing challenges like traffic management, vehicle safety, and sustainability. IoT is not just a technological trend, it is a paradigm shift that promises to reshape the way we move from place to place. The key components and architecture of IoT lay the groundwork for a deeper exploration of its application in transportation, which will be the focus of the chapters that follow.

10.3 SMART TRANSPORTATION

In an era where technology continues to redefine the way we interact with the world, the transformation of transportation into "smart transportation" represents a significant step forward. This chapter delves into the evolution of conventional transportation into a smart and interconnected system and explores the pivotal role played by the IoT in shaping this revolution [12–16].

10.3.1 EVOLUTION INTO SMART TRANSPORTATION

Transportation, once a primarily mechanical and manual system, has undergone a remarkable evolution over the past few decades. The conventional model of transportation, which largely relied on stand-alone vehicles and manual traffic management, faced a multitude of challenges.

- *Traffic Congestion:* Rapid urbanization led to overcrowded road networks, causing traffic jams, increased commute times, and economic losses.
- *Safety Concerns:* High rates of accidents, injuries, and fatalities on the roads prompted concerns about road safety. The environmental consequences of conventional transportation, such as greenhouse gas emissions and pollution, became global concerns.
- *Efficiency:* The inefficiency of resource allocation, including fuel, time, and infrastructure, raised questions about the sustainability of traditional transportation systems.

10.3.2 THE BIRTH OF SMART TRANSPORTATION

In response to these challenges, the concept of "smart transportation" emerged. This paradigm shift represents the integration of technology, data, and real-time decision-making to create a more efficient, -safe, and sustainable transportation system [17–20]. Smart transportation relies on data collection and analysis to optimize various aspects of transportation. It leverages technology to make informed decisions that enhance the overall transportation experience. For instance, *Vehicle Safety:* V2V communication and collision avoidance systems are integral to smart transportation, significantly reducing the risk of accidents. *Sustainability:* route optimization, eco-friendly transportation options, and efficient resource allocation help reduce the environmental footprint of transportation.

10.3.3 IoT's ROLE IN CREATING SMART TRANSPORT SYSTEMS

Here is how IoT contributes to the creation of intelligent transport systems:

- *Real-time Data Collection:* IoT sensors placed on vehicles, roads, and infrastructure collect an immense amount of data in real time. For instance, sensors on vehicles monitor speed, location, and even driver behavior. Roadside sensors track traffic density, weather conditions, and road surface quality. This real-time data forms the backbone of smart transportation systems, allowing for up-to-the-minute decision-making.
- *Connectivity and Communication:* IoT enables seamless communication among various elements of the transportation system. Vehicles can communicate with other vehicles (V2V) and with infrastructure (V2I) through cellular networks. For example, if one vehicle detects an obstacle or hazardous road conditions, it can transmit this information to nearby vehicles, allowing them to adjust their routes or driving behaviors accordingly. Such real-time communication enhances safety and overall traffic management.

10.4 TRAFFIC MANAGEMENT THROUGH IoT

The efficient management of traffic is a critical challenge in modern urban centers. Traffic congestion not only leads to lost productivity but also adversely affects air quality and the quality of life for urban residents. This chapter explores the role of the IoT in revolutionizing traffic management by providing real-time traffic monitoring, implementing congestion reduction strategies, and presenting case studies that demonstrate the tangible impact of IoT in managing traffic.

10.4.1 REAL-TIME TRAFFIC MONITORING

The advent of real-time traffic monitoring marks a substantial advancement in comprehending and regulating traffic flows in urban environments. Unlike traditional traffic management systems that depend on static data and predictions derived from historical patterns, IoT has ushered in an era of live data collection and analysis. This capability empowers authorities to make informed decisions in real time, optimizing traffic flow dynamically as it unfolds.

10.4.2 IoT SENSORS FOR TRAFFIC MONITORING

IoT sensors are at the heart of real-time traffic monitoring. These sensors are strategically placed on roadways, intersections, and even in vehicles, continuously collecting data on traffic conditions. These sensors come in various forms, including Traffic Cameras: These cameras provide visual data on traffic conditions, offering a live feed to traffic management centers. This visual data allows authorities to monitor congestion, accidents, and other disruptions.

10.4.3 DATA FUSION AND ANALYSIS

Real-time traffic data from these sensors is collected and processed in traffic management centers. Advanced data fusion techniques are employed to combine data from various sources, creating a comprehensive picture of traffic conditions. The real-time data is then analyzed to identify congestion, accidents, and other issues that may disrupt traffic flow.

10.4.4 CONGESTION REDUCTION STRATEGIES

The implementation of congestion reduction strategies that help alleviate traffic issues is significantly enhanced by IoT technology. One highly effective method for reducing congestion is through the use of adaptive traffic signal control systems. Traditional traffic signal timing often operates on fixed schedules, which can exacerbate congestion during peak hours. For example, *Green Wave Systems:* these systems coordinate traffic signals to create a “green wave” that allows a platoon of vehicles to travel through a series of intersections without stopping. Traffic signals can dynamically change their timing to respond to varying traffic volumes. This minimizes unnecessary stops and starts, reducing congestion.

10.4.5 PUBLIC TRANSPORTATION INTEGRATION

IoT can also integrate public transportation systems into congestion reduction strategies. By providing real-time data on the locations and schedules of buses and trains, commuters can make informed decisions about using public transportation rather than contributing to road congestion.

10.4.6 CASE STUDIES IN TRAFFIC MANAGEMENT

Real-world case studies provide concrete evidence of the effectiveness of IoT in traffic management. Here are a few notable examples that highlight the impact of IoT on congestion reduction and enhanced traffic management:

10.4.6.1 LOS ANGELES, CALIFORNIA

Los Angeles, known for its notorious traffic congestion, has implemented an advanced traffic management system called Adaptive Traffic Signal Control (ATSAC). ATSAC uses a network of cameras and sensors to monitor traffic flow and adjust signal timings in real-time. The system has led to significant reductions in travel times and fuel consumption.

10.4.6.2 SINGAPORE

Singapore's comprehensive traffic management system, known as the Singapore Intelligent Transport System, utilizes IoT technology to monitor traffic conditions, predict congestion, and provide real-time traffic information to drivers. The city–state has seen substantial improvements in traffic flow and reduced traffic-related emissions.

10.4.6.3 STOCKHOLM, SWEDEN

Stockholm implemented a congestion pricing system utilizing IoT technology to gather tolls according to the location of vehicles and the time of day. This innovative approach has successfully reduced traffic congestion and improved air quality in the city.

10.4.6.4 SAN FRANCISCO, CALIFORNIA

San Francisco has deployed an IoT-driven parking management system that uses sensors to monitor parking space availability in real time. This data is then shared with drivers *via* mobile apps, reducing the time spent searching for parking and mitigating congestion caused by circling for parking spaces.

These case studies demonstrate the real-world impact of IoT on traffic management, ranging from adaptive traffic signal control to congestion pricing and parking management. They emphasize the potential of IoT to reduce congestion, enhance traffic flow, and make transportation more efficient and environmentally friendly [21].

10.5 ENHANCING VEHICLE SAFETY

Ensuring the safety of all road users is a top priority in the transportation sector. Given the high occurrence of accidents and fatalities each year, there is a pressing demand for innovative solutions that can mitigate the risk of accidents and improve overall vehicle safety. In this chapter, we delve into the pivotal role played by the IoT in enhancing vehicle safety. We specifically explore the contributions of V2V communication, collision avoidance systems, and the broader safety implications of IoT.

10.5.1 VEHICLE-TO-VEHICLE COMMUNICATION

One of the most significant advancements in vehicle safety is V2V communication. V2V technology leverages IoT to enable vehicles to communicate with each other in real time, sharing crucial information about their speed, location, and heading. This transformative innovation has the potential to drastically reduce accidents and improve road safety.

10.5.1.1 HOW V2V COMMUNICATION WORKS

V2V communication hinges on Dedicated Short-Range Communication technology, functioning within a designated 5.9 GHz radio frequency band. Vehicles integrated with V2V technology consistently transmit Basic Safety Messages (BSMs) containing crucial details like speed, location, acceleration, and brake status. Nearby vehicles intercept these BSMs, utilizing the information to evaluate potential collision risks.

10.5.1.2 COLLISION AVOIDANCE AND SAFETY APPLICATIONS

The primary purpose of V2V communication is to enable advanced safety applications that can help drivers avoid accidents. Here are some key collision avoidance applications:

- *Intersection Movement Assist (IMA):* IMA provides warnings to drivers when it detects that a collision may occur at intersections. It can warn drivers about vehicles approaching from blind spots, even when they are not visible to the driver.
- *Forward Collision Warning (FCW):* FCW systems use V2V data to assess the relative speed and distance between vehicles. If a potential collision is detected, the driver is alerted through visual and audible warnings.
- *Blind Spot Warning (BSW):* BSW systems utilize V2V information to monitor blind spots and provide alerts when another vehicle is approaching a neighboring lane.
- *Lane Change Warning (LCW):* LCW systems use V2V communication to detect vehicles in adjacent lanes and warn the driver about unsafe lane changes.
- *Emergency Electronic Brake Lights (EEBL):* EEBL applications use V2V data to detect hard braking events in nearby vehicles and send alerts to other vehicles. This can help drivers react more quickly to sudden stops.

10.5.2 COLLISION AVOIDANCE SYSTEMS

In addition to V2V communication, collision avoidance systems are another crucial component of vehicle safety enhanced by IoT. These systems leverage a combination of sensors, cameras, and IoT technology to detect and respond to potential collisions, providing an extra layer of safety for drivers.

10.5.2.1 TYPES OF COLLISION AVOIDANCE SYSTEMS

Collision avoidance systems encompass various technologies designed to prevent accidents and reduce the severity of collisions. Here are some of the most prominent types:

- *Forward Collision Warning (FCW)*: FCW systems use sensors to monitor the distance and relative speed between your vehicle and the one in front of you. If a potential collision is detected, the system alerts the driver.
- *Automatic Emergency Braking (AEB):* AEB takes FCW a step further. If the driver doesn't respond to the FCW warning, AEB can automatically apply the brakes to mitigate or prevent a collision.
- *Rear-View Cameras:* These cameras provide a view of the area behind the vehicle when in reverse, making it easier for drivers to spot obstacles and pedestrians.

10.5.3 THE ROLE OF IoT IN COLLISION AVOIDANCE SYSTEMS

IoT plays a vital role in these Collision Avoidance Systems by connecting various components and enabling real-time data analysis. Here is how IoT technology enhances these systems:

- *Data collection and analysis:* Sensors, cameras, and other components collect data about the vehicle's surroundings. IoT technology facilitates the seamless transmission of this data to onboard computers for analysis.
- *Real-time processing:* IoT allows for real-time processing of data. When a potential collision risk is detected, the system can respond rapidly, providing alerts to the driver or activating emergency braking if necessary.
- *Connectivity:* IoT connectivity extends beyond the vehicle itself. These systems can communicate with other vehicles through V2V communication, further enhancing their ability to detect and respond to potential collisions [22].
- *Data sharing:* In some cases, data from collision avoidance systems can be shared with traffic management centers to improve overall road safety and traffic flow. For example, if a vehicle's system detects

a hazard, it can transmit this information to the traffic management center, which can then relay warnings to other drivers in the area.

10.5.4 SAFETY IMPACT OF IoT

The safety impact of IoT in the realm of vehicle safety is not confined to individual vehicles alone. It extends to broader aspects of transportation safety and even public health. Here is a detailed exploration of the safety impact of IoT:

10.5.4.1 ROAD SAFETY ADVANCEMENTS

IoT technology has ushered in significant advancements in road safety. By facilitating V2V communication and Collision Avoidance Systems, IoT plays a pivotal role in preventing accidents and reducing the severity of collisions. These safety features are not only beneficial for individual drivers but also for pedestrians and cyclists who share the road. The result is fewer injuries and fatalities, as IoT systems provide timely warnings and assist drivers in making safer decisions.

10.5.4.2 TRAFFIC MANAGEMENT AND EMERGENCY RESPONSE

IoT's role in traffic management extends to more than just congestion reduction. Real-time traffic monitoring, often facilitated by IoT sensors, enables faster detection of accidents and incidents. This information is shared with traffic management centers and emergency services, allowing them to respond promptly. Emergency services can be dispatched more quickly to accident scenes, potentially saving lives. Furthermore, IoT-based navigation apps can provide drivers with alternative routes in case of road closures or accidents, minimizing delays and detours.

10.5.4.3 PUBLIC HEALTH AND AIR QUALITY

IoT's impact on safety extends to public health. Traffic congestion and accidents contribute to air pollution, which has adverse health effects, particularly in urban areas. By optimizing traffic flow and reducing the frequency of accidents, IoT indirectly contributes to cleaner air and better public health. Reductions in fuel consumption due to smoother traffic flow also lead to lower greenhouse gas emissions, further promoting environmental and public health benefits.

10.5.4.4 FUTURE POSSIBILITIES

As IoT technology continues to evolve, so do the possibilities for enhancing vehicle safety. Future developments may include more advanced collision avoidance systems, such as those that can detect and respond to potential accidents involving pedestrians or bicyclists. Autonomous vehicles, which rely heavily on IoT technology, have the potential to revolutionize safety further by eliminating the factor of human error in accidents.

10.6 SUSTAINABILITY AND IoT IN TRANSPORTATION

As the world grapples with the environmental consequences of traditional transportation systems, pursuing sustainability within the transportation sector has become a paramount global goal. This chapter delves into the transformative potential of the IoT in achieving sustainable transportation objectives. We explore the core components of sustainable transportation goals, the crucial role that IoT plays in contributing to sustainability, and the manifold environmental benefits of adopting IoT in transportation systems [23–25].

10.6.1 SUSTAINABLE TRANSPORTATION GOALS

Sustainability in transportation is a complex concept that encompasses various objectives and principles with the overarching aim of mitigating the negative environmental, social, and economic impacts of transportation systems. In this context, we elaborate on the fundamental components of sustainable transportation goals:

10.6.1.1 ENHANCE ENERGY EFFICIENCY

Another core element of sustainability in transportation is the enhancement of energy efficiency. Traditional vehicles often operate with a significant amount of energy waste, particularly in urban settings with frequent stops and starts. Sustainable transportation goals prioritize the development of energy-efficient vehicles and the optimization of transportation networks to reduce fuel consumption and reliance on nonrenewable energy sources.

10.6.1.2 ENHANCE AIR QUALITY

Air quality is a significant concern in urban areas, where transportation emissions contribute to the presence of pollutants and particulate matter in the atmosphere. Sustainable transportation goals strive to improve air quality by reducing emissions from vehicles and promoting cleaner transportation options, ultimately enhancing public health and well-being.

10.6.2 IoT's CONTRIBUTION TO SUSTAINABILITY

The IoT is a game-changing technology that plays a pivotal role in achieving sustainable transportation goals. Here, we delve into the manifold ways in which IoT contributes to sustainability within the transportation sector:

10.6.2.1 ROUTE OPTIMIZATION

IoT technology enables real-time data collection and analysis that can be used to optimize transportation routes. By gathering data on traffic conditions, road closures, and other factors, IoT systems can provide drivers with information on the most efficient and least congested routes. This not only reduces travel time but also minimizes fuel consumption and emissions.

10.6.2.2 ECO-FRIENDLY TRANSPORTATION ALTERNATIVES

IoT facilitates the growth of eco-friendly transportation alternatives. Through ride-sharing apps and vehicle-sharing platforms, IoT can match riders with drivers and encourage carpooling, reducing the number of vehicles on the road. Additionally, IoT-driven mobility services can promote the use of electric vehicles and bicycles, contributing to sustainability by reducing emissions and reliance on fossil fuels.

10.6.2.3 ENERGY-EFFICIENT VEHICLE OPERATION

IoT can enhance the energy efficiency of vehicle operation. Sensors and data analytics can provide real-time feedback to drivers on their driving habits, encouraging fuel-efficient driving practices. Furthermore, IoT can enable the remote monitoring and management of vehicle fleets, optimizing maintenance schedules and ensuring vehicles operate at peak efficiency.

10.6.2.4 ENVIRONMENTAL MONITORING

IoT's capabilities extend to environmental monitoring. Sensors can measure air quality, noise pollution, and other environmental factors. This data is invaluable for urban planners and policymakers seeking to create sustainable transportation systems that prioritize environmental quality.

10.6.3 ENVIRONMENTAL BENEFITS

The adoption of IoT in transportation yields a myriad of environmental benefits, which extend far beyond mere efficiency gains. Here, we illuminate the specific advantages of IoT in reducing the environmental footprint of transportation.

10.6.3.1 REDUCTION IN GREENHOUSE GAS EMISSIONS

Perhaps the most significant environmental benefit of IoT in transportation is the reduction of greenhouse gas emissions. By optimizing routes, promoting eco-friendly transportation alternatives, and enhancing energy efficiency, IoT contributes to a substantial decrease in the carbon footprint of the transportation sector. This is vital in addressing global climate change concerns.

10.6.3.2 IMPROVED AIR QUALITY

IoT-driven traffic management and congestion reduction measures not only cut emissions but also lead to improved air quality in urban areas. Reductions in pollution and particulate matter benefit public health.

10.6.3.3 NOISE POLLUTION MITIGATION

IoT sensors can monitor noise levels in urban environments. By optimizing traffic flow and reducing congestion, IoT contributes to quieter and more peaceful urban areas.

10.6.3.4 EFFICIENT RESOURCE UTILIZATION

IoT technology enhances resource efficiency by optimizing the use of transportation infrastructure. Reduced congestion, better traffic flow, and efficient allocation of resources such as fuel and time all contribute to more sustainable resource utilization.

10.6.3.5 BICYCLES, ELECTRIC VEHICLES, AND SHARED MOBILITY

The growth of IoT-driven mobility services encourages the use of bicycles, electric vehicles, and shared mobility platforms. These eco-friendly transportation alternatives reduce the reliance on gasoline-powered vehicles, promote sustainability, and lower emissions.

10.7 CHALLENGES AND CONCERNS IN IoT-ENABLED TRANSPORTATION

The infusion of the IoT into transportation systems presents numerous opportunities, accompanied by its share of challenges and apprehensions.

This chapter delves into the intricacies that emerge at the intersection of IoT and the transportation sector. We explore pivotal issues encompassing data security and privacy, regulatory hurdles, and ethical considerations. This examination shines a light on the crucial factors requiring attention as we navigate the landscape of IoT-enabled transportation.

10.7.1 DATA SECURITY AND PRIVACY ISSUES

The intersection of IoT technology with transportation systems produces extensive data, constituting a valuable resource. Yet, the abundance of this data also gives rise to substantial challenges, particularly in the realms of data security and privacy. In this context, we delve into the intricate web of concerns associated with data security and privacy within the domain of IoT-enabled transportation.

10.7.1.1 DATA SENSITIVITY

One of the primary issues in IoT-enabled transportation is the sensitivity of the data being collected. Information gathered from vehicles and infrastructure can encompass highly personal details, such as travel patterns, routes, destinations, and even the habits of individual travelers. This sensitivity raises concerns about data misuse, unauthorized access, and the potential for privacy breaches.

10.7.1.2 DATA OWNERSHIP AND CONTROL

Determining the ownership and control of IoT-generated transportation data is a complex matter. The data may be collected by vehicle manufacturers, transportation authorities, service providers, or even third-party applications. The lack of clarity regarding data ownership can lead to disputes and challenges in defining responsibilities for data security and privacy.

10.7.1.3 DATA ENCRYPTION AND PROTECTION

Securing the transmission and storage of IoT-generated data is crucial. Encryption and robust security measures are essential to prevent data breaches and unauthorized access. Inadequate data protection measures can expose sensitive information to cyberattacks and other security threats.

10.7.1.4 USER CONSENT AND TRANSPARENCY

Ensuring that users are aware of data collection practices and obtain their consent for data usage is a fundamental aspect of data privacy. Many IoT-enabled transportation services rely on user data for various purposes, such as traffic monitoring, route optimization, and advertising. It is vital to establish transparent practices for obtaining user consent and informing them about how their data will be used.

10.7.1.5 DATA SHARING AND THIRD-PARTY ACCESS

IoT in transportation often involves sharing data with third-party applications and services. Sharing data through connected devices brings advantages such as receiving live traffic updates and navigation support. However, it also brings risks. Safeguarding the security of data during sharing is a multifaceted issue that must be addressed.

10.7.2 REGULATORY CHALLENGES

IoT-enabled transportation introduces a host of regulatory challenges that demand careful consideration and resolution. Navigating the complex regulatory landscape is essential to ensure the safe and responsible integration of IoT technology into transportation systems. In this segment, we investigate some of the regulatory challenges linked to the implementation of IoT in transportation.

10.7.2.1 PRIVACY REGULATIONS

Privacy regulations are intertwined with data protection laws but also encompass broader aspects of privacy in transportation. Regulations often dictate how transportation data can be used, who can access it, and the rights of individuals regarding their data. Compliance with these regulations requires a nuanced understanding of privacy principles and robust governance.

10.7.2.2 LIABILITY AND ACCOUNTABILITY

Determining liability in the event of accidents or incidents involving IoT-enabled transportation systems can be particularly challenging. In a connected environment, multiple stakeholders, including vehicle manufacturers, software developers, and infrastructure operators, share responsibility. Regulatory frameworks must establish clear lines of accountability.

10.7.2.3 SAFETY STANDARDS AND CERTIFICATION

Ensuring the safety of IoT-enabled transportation systems necessitates the establishment of safety standards and certification processes. Regulatory agencies must define these standards to guarantee the secure operation of interconnected vehicles and infrastructure.

10.7.2.4 INTERNATIONAL HARMONIZATION

IoT-enabled transportation systems often cross international boundaries. Achieving regulatory consistency and harmonization among different countries is an intricate task, as each nation may have unique regulatory frameworks and standards. Harmonization efforts are essential to facilitate the global deployment of IoT transportation solutions.

10.7.2.5 SPECTRUM AND NETWORK MANAGEMENT

IoT communication relies on radio frequencies, which are finite and regulated resources. Spectrum management is essential to prevent interference and ensure efficient data transmission. Transportation IoT systems require careful coordination to avoid conflicts with other wireless communication systems.

10.7.3 ETHICAL CONSIDERATIONS

In addition to data security and regulatory challenges, ethical considerations play a pivotal role in the adoption of IoT in transportation. The following sections outline the complex ethical questions that emerge as IoT technology becomes increasingly intertwined with transportation systems.

10.7.3.1 DATA OWNERSHIP AND CONTROL

The ethical dilemma of data ownership and control centers on who has the right to own and manage the vast amounts of transportation data generated by IoT systems. As data often pertains to individual behavior and movement, ethical considerations demand that data ownership respect the rights and privacy of the individuals involved.

10.7.3.2 CONSENT AND TRANSPARENCY

Obtaining informed consent from users for data collection and usage is not merely a regulatory requirement; it is also an ethical imperative. Transparency in data practices is a fundamental ethical principle, as users should be fully aware of how their data is used and have the autonomy to grant or deny consent.

10.7.3.3 EQUITY AND ACCESSIBILITY

Ensuring equity and accessibility is an ethical responsibility in IoT-enabled transportation. Transportation solutions should be designed to benefit all members of society, regardless of their socioeconomic status or technological literacy. Ethical considerations call for the avoidance of digital divides and ensuring that the benefits of IoT are inclusive.

10.7.3.4 ACCOUNTABILITY AND RESPONSIBILITY

Determining responsibility in IoT-enabled transportation accidents is a complex ethical issue. As interconnected systems introduce multiple stakeholders, clarifying accountability becomes essential. Ethical frameworks must address questions of responsibility and fairness when accidents or incidents occur.

10.7.3.5 DATA MINIMIZATION AND PURPOSE LIMITATION

Ethical data practices demand that data collected for transportation purposes be minimized and used solely for the intended purposes. Gathering excessive data or using it for unrelated purposes may infringe on privacy and ethical standards.

10.7.3.6 ENVIRONMENTAL IMPACT

IoT-enabled transportation has both positive and negative environmental impacts. Ethical considerations require that IoT solutions prioritize environmental sustainability and minimize harm to the ecosystem, considering the broader ethical implications of transportation choices.

In conclusion, the integration of IoT in transportation introduces a complex web of challenges and concerns. Data security and privacy issues, regulatory challenges, and ethical considerations are pivotal factors that demand careful attention and resolution. Addressing these challenges is essential to ensure the responsible and sustainable integration of IoT technology into transportation systems, while also respecting privacy and ethical standards.

10.8 FUTURE TRENDS IN IoT TRANSPORTATION

As IoT technology continues to evolve and reshape the transportation landscape, the future of IoT-enabled transportation holds immense promise and potential. This chapter is dedicated to exploring the forthcoming trends and advancements in IoT transportation, including emerging technologies, predictions, and possibilities that are set to redefine the way we move and interact with the world [26–30].

10.8.1 EMERGING TECHNOLOGIES

The future of IoT transportation is characterized by an array of emerging technologies that have the power to revolutionize mobility, enhance safety, and improve sustainability. Here, we delve into some of the most noteworthy emerging technologies that are poised to shape the future of transportation.

10.8.1.1 5G CONNECTIVITY

The deployment of 5G networks is poised to revolutionize connectivity in transportation. 5G presents unprecedented speed and low latency, making it well-suited for real-time communication among vehicles, infrastructure, and the cloud. This technology allows vehicles to exchange data rapidly and reliably, facilitating a diverse array of IoT applications, including V2V communication, traffic management, and entertainment services for passengers.

10.8.1.2 MOBILITY AS A SERVICE

Mobility as a Service (MaaS) represents a burgeoning concept aimed at revolutionizing transportation into a unified and integrated experience. MaaS platforms amalgamate diverse transportation modes, including public transit, ride-sharing, and bike-sharing, into a singular and user-friendly service. Through a single app, passengers can plan, book, and pay for their entire journey, diminishing the reliance on car ownership and fostering sustainable transportation practices.

10.8.1.3 ELECTRIC AND HYDROGEN-POWERED VEHICLES

The transition to electric and hydrogen-powered vehicles is gaining momentum as the world seeks to reduce greenhouse gas emissions and reliance on fossil fuels. Electric vehicles (EVs) and fuel cell vehicles powered by hydrogen are becoming increasingly accessible and affordable.

IoT technology will play a significant role in EV charging infrastructure and battery management, making these vehicles more practical and environmentally friendly.

10.8.1.4 ADVANCED TRAFFIC MANAGEMENT SYSTEMS

The integration of IoT technology is driving the evolution of traffic management systems. Advanced systems leverage real-time data from IoT sensors to optimize traffic flow, alleviate congestion, and enhance safety. Machine learning and AI algorithms play a pivotal role, enabling these systems to predict traffic patterns and dynamically adjust traffic signals for improved efficiency.

10.8.1.5 DRONE AND AERIAL TRANSPORTATION

The use of drones and aerial transportation is a fascinating emerging technology in IoT transportation. Drones have the potential to transport goods and people efficiently, especially in urban areas with traffic congestion. They could also be used for emergency medical services, surveillance, and remote sensing applications.

10.8.2 PREDICTIONS AND POSSIBILITIES

The future of IoT transportation is filled with predictions and possibilities that can reshape the way we move and interact with our environment. Here, we explore some of the key forecasts and opportunities that lie ahead.

10.8.2.1 REDUCED TRAFFIC CONGESTION

With the adoption of autonomous vehicles, improved traffic management, and the growth of MaaS, the future holds the potential for significantly reduced traffic congestion in urban areas. Commutes may become smoother and less time-consuming, enhancing the quality of life for urban dwellers.

10.8.2.2 IMPROVED ACCESSIBILITY

MaaS and IoT transportation solutions have the potential to make transportation more accessible to a wider range of individuals, including those with disabilities or limited mobility. The future could see more inclusive and user-friendly transportation services.

10.8.2.3 PERSONALIZED TRANSPORTATION SERVICES

IoT-enabled transportation can offer personalized and on-demand transportation services. Users can tailor their travel experiences to suit their preferences, whether that means choosing autonomous vehicles for convenience or shared rides for cost-effectiveness.

10.8.2.4 ECONOMIC OPPORTUNITIES

The growth of IoT transportation technology presents economic opportunities in various sectors, from the development and manufacturing of autonomous vehicles to the creation of new mobility services and data analytics companies. The transportation industry is undergoing a transformation that will lead to new business models and job opportunities.

10.8.2.5 URBAN PLANNING AND DESIGN

The integration of IoT technology in transportation will influence urban planning and design. Cities may evolve to prioritize pedestrian-friendly environments, efficient public transit, and sustainable infrastructure. IoT data will play a crucial role in shaping cities that are more livable and environmentally friendly.

10.9 CASE STUDIES IN IoT TRANSPORTATION

The real-world application of IoT technology in transportation is replete with examples that illustrate the transformative impact of interconnected systems. This chapter delves into case studies of IoT implementation, offering insight into successful deployments, innovative solutions, and the lessons learned from these experiences.

10.9.1 REAL-WORLD EXAMPLES OF IoT IMPLEMENTATION

10.9.1.1 SMART CITY INITIATIVES: SINGAPORE'S SMART MOBILITY SYSTEM

Singapore has emerged as a beacon of smart urban development, leveraging IoT technology to create efficient and sustainable transportation systems. One standout initiative is the Smart Mobility System, which combines various IoT components to optimize the flow of vehicles and commuters.

10.9.1.2 REAL-TIME TRAFFIC MANAGEMENT

Singapore employs an extensive network of sensors and cameras that monitor traffic conditions in real time. This data feeds into a central control center, allowing traffic managers to make data-driven decisions regarding traffic signal timing and lane management. This approach minimizes congestion and reduces travel time for commuters.

10.9.1.3 ELECTRONIC ROAD PRICING

Electronic road pricing is a toll system that adjusts charges based on traffic conditions. IoT sensors communicate with vehicles' onboard units to calculate and collect tolls, enabling dynamic pricing to discourage travel during peak hours and on congested routes. This system not only generates revenue but also mitigates traffic congestion.

10.9.1.4 PARKING AVAILABILITY

IoT sensors in parking lots and street-side parking spaces monitor occupancy. This data is made accessible to drivers through mobile apps, enabling them to locate available parking spaces quickly. As a result, drivers spend less time circling for parking, reducing both traffic and emissions.

10.9.2 V2V AND V2I COMMUNICATION: VOLVO'S CONNECTED SAFETY TECHNOLOGY

Volvo's commitment to road safety led to the development of its Connected Safety technology, which relies on V2V and V2I communication to enhance vehicle safety and prevent accidents.

10.9.2.1 V2V COMMUNICATION

Volvo vehicles equipped with V2V technology can communicate with each other, sharing real-time data about their speed, position, and direction. When one vehicle detects a potential collision risk, it can send warnings to other nearby vehicles. This instantaneous exchange of information fosters proactive collision avoidance and a safer driving environment.

10.9.2.2 V2I COMMUNICATION

In addition to V2V communication, Volvo's technology interacts with infrastructure (V2I). For example, when a vehicle approaches a connected traffic signal, it can receive information about the signal's timing, enabling the driver to adjust speed and reduce the likelihood of running a red light.

10.9.3 SUCCESS STORIES AND LESSONS LEARNED

10.9.3.1 SUCCESS STORY: MADRID'S SUSTAINABLE URBAN MOBILITY PLAN

Madrid, Spain, implemented a Sustainable Urban Mobility Plan (SUMP) to address traffic congestion, air quality, and sustainability. The plan's success is rooted in comprehensive IoT integration and several key features:

- *Integrated Public Transport:* Madrid deployed a unified fare system for all public transportation modes, including buses, trams, and the metro. Commuters can use a single card or mobile app to access and pay for multiple modes, promoting the use of public transport.
- *Real-time Information:* IoT technology provides real-time information on public transport schedules, route options, and traffic conditions. Commuters can access this information through mobile apps, making informed choices about their journeys.
- *Eco-friendly Initiatives:* Madrid introduced low-emission zones, banning high-pollution vehicles from certain areas. IoT sensors monitor vehicle emissions and enforce compliance. The initiative has led to improved air quality and reduced traffic congestion.
- *Lesson Learned:* Madrid's success lies in its holistic approach to transportation, combining integrated public transport, real-time information, and sustainability initiatives. The lesson here is that comprehensive planning and integration of IoT solutions can yield tangible mobility and environmental quality benefits.

10.9.3.2 SUCCESS STORY: MAAS IN HELSINKI, FINLAND

Helsinki, Finland, has embraced the concept of mobility as a Service (MaaS) to create a seamless and user-centric transportation system. Key elements of Helsinki's MaaS success include:

- *MaaS Applications:* Helsinki developed MaaS applications that allow users to plan and pay for their journeys using a single platform. Commuters can combine various modes of transport, from buses and trams to taxis and bike-sharing, into a single, integrated itinerary.
- *Dynamic Pricing:* The pricing of journeys on the MaaS platform is dynamic, factoring in the selected modes of transport and real-time demand. This encourages the use of less congested modes during peak hours, reducing traffic.
- *Accessibility:* The MaaS platform prioritizes accessibility and inclusivity. It offers options for individuals with disabilities and promotes the use of public transport and eco-friendly alternatives.
- *Lesson Learned:* Helsinki's MaaS success demonstrates the power of user-centric transportation solutions. By placing users at the center and offering flexibility and accessibility, the city has created

a model that promotes sustainable mobility while reducing traffic congestion.

10.9.3.3 SUCCESS STORY: PORTLAND, OREGON'S ECO-COUNTER SYSTEM

Portland, Oregon, implemented the Eco-Counter system, an IoT-based solution that promotes cycling as a sustainable mode of transportation.

- *Eco-Counters:* Portland installed eco-counters at various locations, equipped with sensors that track the number of cyclists using dedicated bike lanes. The data collected from these counters provides insights into cycling patterns and trends.
- *Real-time Data:* The eco-counter system offers real-time data on cyclist volumes and their use of bike lanes. This data is made available to the public, allowing cyclists to choose less congested routes and making urban planning more data-driven.
- *Lesson Learned:* Portland's success in promoting cycling through IoT technology underscores the importance of data-driven decision-making and infrastructure improvements. Real-time data accessibility benefits both cyclists and urban planners, creating a win-win situation for sustainable transportation.

In conclusion, these case studies of IoT implementation in transportation offer valuable insights into the potential of interconnected systems. Success stories from cities like Singapore, Madrid, Helsinki, and Portland demonstrate the transformative power of IoT technology in optimizing transportation, enhancing safety, and promoting sustainability. The lessons learned from these examples underscore the significance of comprehensive planning, user-centric solutions, and data-driven decision-making in the pursuit of a more efficient, safe, and sustainable transportation future.

10.10 CONCLUSION

The journey through the intersection of the IoT and transportation has been nothing short of transformative, unveiling a future filled with possibilities, innovations, and renewed hope for efficient, sustainable, and safe mobility.

This concluding section summarizes the key takeaways from our exploration and underscores IoT's transformative impact on transportation.

10.10.1 RECAP OF KEY TAKEAWAYS

Throughout this chapter, we have unearthed several essential takeaways that encapsulate the essence of IoT's role in transportation. Let us revisit these insights:

- *IoT as a Disruptive Force:* IoT has emerged as a disruptive force across multiple sectors, with transportation being no exception. It has redefined how we conceive, manage, and interact with our transportation systems.
- *Multifaceted Transportation Challenges:* The transportation industry faces a spectrum of challenges, including traffic congestion, road safety concerns, and environmental repercussions. These issues have far-reaching implications for productivity, public health, and the environment.
- *Smart Transportation through IoT:* IoT is not just about technology, it symbolizes the era of "smart transportation." By leveraging sensors, real-time data communication, and sophisticated analytics, IoT bestows intelligence, efficiency, and sustainability upon transportation systems.
- *IoT's Role in Traffic Management:* One of IoT's pivotal roles in transportation is traffic management. Through data-driven decision-making, IoT mitigates traffic congestion, resulting in smoother commutes and optimized road networks. This is essential for improved productivity and air quality.
- *Enhancing Vehicle Safety:* IoT technologies enhance vehicle safety by facilitating communication between vehicles (V2V) and deploying collision avoidance systems. This not only minimizes accidents but also contributes to safer roadways.
- *Sustainability through IoT:* IoT transforms transportation into a greener, more eco-conscious enterprise. IoT's contributions to sustainability include route optimization, emissions reduction, and the promotion of eco-friendly transportation alternatives.
- *Future Trends and Possibilities:* The future of IoT transportation is brimming with potential. Emerging technologies like autonomous

vehicles, 5G connectivity, MaaS, and electric vehicles are set to redefine how we move. Predictions encompass reduced traffic congestion, enhanced safety, greater sustainability, improved accessibility, and new economic opportunities.

10.10.2 IoT's TRANSFORMATIVE IMPACT ON TRANSPORTATION

The impact of IoT on transportation is nothing short of transformative. As we conclude this exploration, let us delve into the profound changes it has wrought.

- *Efficiency and Productivity:* IoT-enabled transportation has unleashed newfound efficiency in travel. Real-time data analytics, route optimization, and traffic management have reduced travel time, minimized congestion, and, in turn, increased productivity. Commuters spend less time stuck in traffic and more time engaging in meaningful activities.
- *Safety and Security:* Safety is paramount in transportation, and IoT technology has proven to be a stalwart ally in this regard. The adoption of V2V communication and collision avoidance systems has significantly reduced accidents and fatalities. The roads have become safer, and the fear of accidents has diminished.
- *Sustainability and Environmental Responsibility:* The environmental repercussions of transportation have been a growing global concern, primarily due to greenhouse gas emissions. IoT has played a pivotal role in addressing this concern. By optimizing routes, promoting eco-friendly alternatives, and reducing emissions, IoT has contributed to a more environmentally responsible transportation ecosystem.
- *Enhanced User Experience:* IoT has ushered in an era of user-centric transportation. From real-time information on public transport schedules to the convenience of MaaS platforms, the focus has shifted toward providing an enhanced user experience. Commuters can make informed choices, select sustainable transportation options, and enjoy seamless journeys.
- *Economic Growth and Innovation:* The integration of IoT in transportation has triggered economic growth and innovation. New business models have emerged, from autonomous vehicle development to data analytics services. Job opportunities have expanded, and innovation has become a driving force in the transportation sector.

- *Improved Urban Planning:* The influence of IoT technology has extended to urban planning and design. Smart cities are prioritizing pedestrian-friendly environments, efficient public transit, and sustainable infrastructure. The data-driven insights from IoT have paved the way for improved urban living.

KEYWORDS

- **Internet of Things**
- **smart transportation**
- **traffic management**
- **vehicle safety**
- **sustainable mobility**

REFERENCES

1. Gubbi, J., Buyya, R., Marusic, S., & Palaniswami, M. Internet of Things (IoT): A vision, architectural elements, and future directions. *Future Generation Computer Systems* **2013**, *29*(7) 1645–1660.
2. Atzori, L., Iera, A., & Morabito, G. The Internet of Things: A survey. *Computer Networks* **2010**, *54*(15), 2787–2805.
3. Zanella, A., Bui, N., Castellani, A., Vangelista, L., & Zorzi, M. Internet of Things for smart cities. *IEEE Internet of Things Journal* **2014**, *1*(1), 22–32.
4. Kortuem, G., Kawsar, F., Fitton, D., Sundramoorthy, V., & Alford, D. Smart objects as building blocks for the Internet of Things. *IEEE Internet Computing* **2010,** *14*(1), 44–51.
5. Perera, C., Liu, C. H., Jayawardena, S., & Chen, M. A survey on Internet of Things from an industrial market perspective. *IEEE Access* **2014**, *2*, 1660–1679.
6. Tomic, S., Holz, C., Sonntag, D., & Krco, S. The Internet of Things in the industry 4.0 environment. *Procedia CIRP* **2014**, *16*, 180–185.
7. Bhattacharya, S., & Jana, P. K. IoT-based smart cities: A survey. *IETE Technical Review* **2017**, *34*(1), 1–9.
8. Han, R., & Shi, L. Smart grid and the Internet of Things: A review. *International Journal of Distributed Sensor Networks* **2018**, *14*(10), 1550147718803217.
9. Wang, D., He, S., Jin, Q., & Vasilakos, A. V. Deep learning for smart cities. *Information Fusion* **2018**, *42,* 38–53.
10. Bandyopadhyay, D., & Sen, J. Internet of things: Applications and challenges in technology and standardization. *Wireless Personal Communications* **2011**, *58*(1), 49–69.

11. Farooq, M. S., & Kunzel, K. Internet of Things, smart cities and urban planning. In: *Proceeding of Internet of Things and Big Data Technologies for Next Generation Healthcare*, **2015**, pp. 37–65. Springer.
12. Sun, H., & De Florio, V. IoT platform for sustainable urban development in smart cities. In: *Proceedings of ACM International Joint Conference on Pervasive and Ubiquitous Computing*, **2016**, pp. 471–474.
13. Atzori, L., Antonio, I., Morabito, G., & Nitti, M. The social Internet of Things (SIoT)—when social networks meet the Internet of Things: Concept, architecture and network characterization. *Computer Networks* **2012** *56*(16), 3594–3608.
14. Zheng, Z., Zheng, Y., Xie, X., & Yang, Q. Collaborative filtering meets mobile recommendation: A user-centered approach. *ACM Transactions on Intelligent Systems and Technology (TIST)* **2015**, *5*(4), 1–33.
15. Ma, X., Yang, H., Liu, X., Yang, Y., & Guan, L. Wireless communications and networking for smart grid: An overview. *IEEE Transactions on Industrial Electronics* **2013**, *60*(4), 1596–1607.
16. Tampuu, A., Matskin, M., Bell, M., & Laanpere, M. Smart citie. In: *Handbook of Research on E-Services in the Public Sector: E-Government Strategies and Advancements* **2014**, pp. 738–764.
17. Agarwal, A., & Shafique, M. A survey of security issues in wireless sensor networks, arXiv preprint arXiv:1803.05068, 2018.
18. Atzori, L., Iera, A., & Morabito, G. From 'smart objects' to 'social objects': The next evolutionary step of the Internet of Things. *IEEE Communications Magazine* **2012**, *50*(3), 92–99.
19. Turcanu, I., Pekhteryev, G., Jornet, J. M., & Akyildiz, I. F. Terahertz-band communication and imaging: A new horizon for wireless communications. *IEEE Transactions on Terahertz Science and Technology* **2018**, *6*(5), 733–740.
20. Botta, A., De Donato, W., Persico, V., & Pescapé, A. Integration of cloud computing and Internet of Things: A survey. *Future Generation Computer Systems* **2016**, *56*, 684–700.
21. Liu, X., Fu, S., Li, Q., & Zhou, J. Privacy-preserving public auditing for secure cloud storage. *IEEE Transactions on Computers* **2015**, *64*(2), 409–417.
22. Katz, G., Gopalakrishnan, V., & Kurakula, P. Software-defined networking: Challenges and research opportunities. *IEEE Communications Surveys & Tutorials* **2017**, *19*(1), 77–104.
23. Chen, X., Kurakula, P., & Gollakota, S. Feasibility and challenges of 60 GHz automotive radar with passive collision detection. *Proceedings of the ACM on Interactive, Mobile, Wearable and Ubiquitous Technologies (IMWUT)* **2019**, *3*(2), 1–28.
24. Tarasov, V., Leppänen, T., Iinatti, J., & Hämäläinen, T. D. A novel decentralized and self-adaptive network architecture for the Internet of Things. *IEEE Transactions on Industrial Informatics* **2018**, *14*(8), 3501–3509.
25. Papageorgiou, A., Diou, C., & Verikoukis, C. IoT for traffic management in smart cities. *Internet of Things: IoT Infrastructures*, **2016**, 285–303.
26. Hou, Y. T., & Liu, Y. T. Big data analytics in traffic and transportation engineering: Emerging research challenges. *Transportation Research Part C: Emerging Technologies* **2017**, *79*, 96–98.

27. Nabeel, M., Alsamhi, S. H., Yaqoob, I., Imran, M., & Guizani, M. IoT-based big data: From data to knowledge and decision. *IEEE Communications Magazine* **2018**, *56*(2), 40–46.
28. Han, J., Shang, J., Zhang, Y., & Yan, R. Smart city development in China: An empirical study. *Sustainable Cities and Society* **2018** (38), 612–624.
29. Chen, Y., Cheng, C. Y., Lin, B. S. P., & Lin, Y. A smart city experiment: The case of Copenhagen Solutions Lab. *Sustainable Cities and Society* **2018**, *38,* 243–254.
30. Raul, P., Andrade, A., & Silva, C. M. The Internet of Things in smart cities: A reality where technology and citizenship can converge for an inclusive city. In: *Proceeding of 10th Iberian Conference on Information Systems and Technologies (CISTI)*, **2015**, pp. 1–6.

CHAPTER 11

Self-Powered IoT Devices: Harnessing Energy for a Connected Future

ABHISHEK KUMAR* and PANKAJ KUMAR KESHRI

School of Electronics and Electrical Engineering, Lovely Professional University, Phagwara, Punjab, India

*Corresponding author

ABSTRACT

The Internet of Things' (IoT) explosive growth has transformed a number of industries by spurring innovation and improving connection. The challenges of making IoT devices efficient, reliable, and environmentally friendly have grown in importance as the number of these devices increases. This chapter examines the idea of self-powered IoT devices, which do not require traditional power sources like batteries by using ambient energy sources including thermal, kinetic, solar, and radio frequency (RF) energy. These devices not only help IoT networks become more sustainable, but they also solve a number of significant drawbacks of conventional power solutions, such as their high maintenance costs, short use, and environmental effects. The chapter explores the basics, advantages, and difficulties of several energy harvesting systems, emphasizing how they facilitate IoT deployment in difficult, enhance scalability, and enable continuous, real-time operations. In order to maximize total power consumption and improve network efficiency, it also covers the incorporation of energy-efficient technology into IoT devices. Examined are the uses of self-powered IoT devices in industries including smart homes, industrial automation, and healthcare, showcasing its revolutionary possibilities. Self-powered devices are essential to the development of

IoT Potential for Green Energy Solutions. Abhishek Kumar, Ajay Roy, & Shamimul Qamar (Eds.)

DOI: 10.1201/9781779643599-11

connected technologies because they solve energy constraints and provide a route to an economical, sustainable IoT environment.

11.1 INTRODUCTION

An innovative concept known as the Internet of Things (IoT) envisions commonplace objects becoming networked and able to exchange data *via* the Internet. The seamless communication and data sharing between devices made possible by this interconnection promote an environment that is more automated and intelligent. Technological developments in connection, sensor technology, and data analytics have propelled the IoT's explosive growth. IoT has impacted many industries, from wearables and smart homes to industrial applications and healthcare systems, offering better user experiences, increased productivity, and better decision-making. IoT applications are constantly growing, which highlights their critical role in reshaping the digital landscape and affecting our interactions with the outside world. Self-powered IoT devices explores the crucial area of energy independence in the IoT [1]. The need for self-sufficient and sustainable devices grows as the IoT expands. This chapter examines a range of energy harvesting systems, such as thermal, kinetic, and solar, and clarifies their workings, limitations, and uses. The conversation also includes radio frequency (RF) energy harvesting and battery-free IoT designs [2, 3]. It is impossible to exaggerate the significance of self-powered devices in overcoming IoT implementation obstacles. Conventional power sources, including batteries, are a common source of power for traditional IoT devices. These sources have drawbacks in terms of longevity, upkeep, and environmental effects. A self-powered device is preferred due to the following features [4–7]:

1. *Sustainability:* Self-sufficient gadgets use thermal, kinetic, or solar energy to extract energy from their surroundings. By reducing reliance on disposable batteries, this sustainable strategy promotes a more sustainable and eco-friendly IoT ecosystem.
2. *Decreased Maintenance:* Changing batteries on a regular basis is one of the key issues with traditional IoT deployments. Because self-powered devices produce their own energy, their operating lifespans are greatly extended, which minimizes the need for frequent interventions and lowers overall maintenance costs.

3. *Remote and Inaccessible Locations:* It might not be feasible to use traditional power sources in remote or inaccessible areas. The range of IoT applications is increased by self-powered devices, which use energy collecting technologies to enable IoT installations in places with sparse or nonexistent power infrastructure.
4. *Enhanced Scalability:* The logistical difficulties associated with maintaining and swapping out the batteries in a large number of devices sometimes impede the scalability of IoT deployments. Self-powered devices increase scalability by lessening the logistical load, making it easier to install IoT networks on a bigger scale.
5. *Continuous Operation:* Self-powered devices have the ability to run continuously, guaranteeing communication and data collecting without interruption. For applications like real-time monitoring, where disruptions might result in the loss of important data, this continuous operation is essential.
6. *Flexibility in Design:* Device design is made flexible by the adaptability of energy collecting technology. More customization and efficiency in IoT solutions are made possible by the ability of devices to be developed to adapt to particular environments and application needs.
7. *Cost-Efficiency:* Although sophisticated technology may result in higher initial prices for self-powered devices, over time, these devices' longer lifespans, lower maintenance requirements, and improved operational efficiency all add to their total cost-effectiveness.
8. *Energy-Efficient IoT Networks:* Energy-efficient technologies are frequently incorporated into self-powered devices, which optimize power consumption and boost the overall effectiveness of IoT networks. In order to create durable and sustainable IoT infrastructures, energy consciousness is essential.

Self-powered device adoption in IoT deployments is consistent with the overarching objectives of cost-effectiveness, sustainability, and enhanced operational efficiency. By addressing the issues with conventional power sources, these gadgets significantly contribute to the development of a more resilient and flexible IoT ecosystem.

11.2 ENERGY HARVESTING SOURCES

Energy harvesting technologies [8] are essential for enabling the independence and sustainability of IoT devices since they are able to capture energy from natural sources as shown in Figure 11.1. Thermal, kinetic, and solar energy are the three main types of energy harvesting. Energy harvesting technologies provide a promising solution to mitigate the challenges associated with conventional power sources in IoT devices. By leveraging solar, kinetic, and thermal energy, IoT devices can operate efficiently, sustainably, and often autonomously, contributing to the advancement of a more energy-efficient and eco-friendly IoT ecosystem [9].

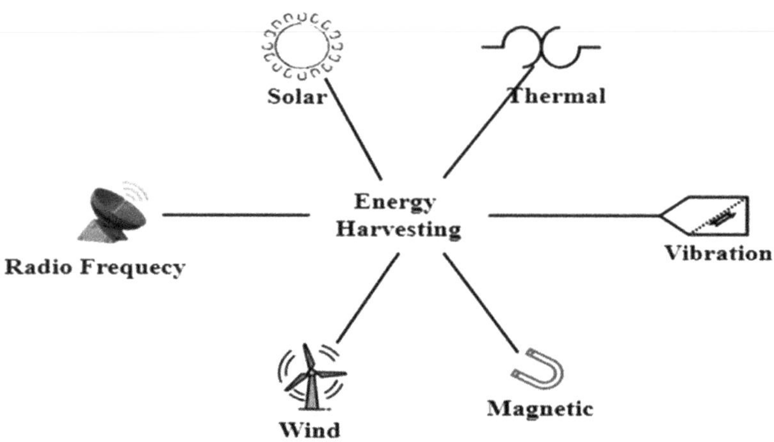

FIGURE 11.1 Energy harvesting sources.

Source: Adapted from Ref. [11].

11.2.1 SOLAR ENERGY

Using photovoltaic cells, solar energy harvesting transforms sunlight into electrical energy. When these cells are exposed to light, they produce a voltage that may be used to create an electric current that powers IoT devices. Numerous IoT products, such as wearable technology, solar-powered streetlights, and outdoor sensors, use solar energy harvesting. However, the presence of sunshine is a need for its efficiency, which restricts its use in indoor or low-light settings. Sunlight emerges as the optimal energy source for harvesting, offering an estimated potential of 1.4 kW/m^2 [10]. Despite the theoretical efficiency of a photovoltaic (PV) cell being 90%, the practical efficiency hovers around 40% [11]. These PV cells exhibit sustained high performance for approximately 25 years, beyond which there is a gradual decline in the generated power.

11.2.2 THERMAL ENERGY

The process of thermal energy harvesting uses thermoelectric generators to produce power based on temperature differentials. These generators take advantage of the Seebeck phenomenon, which occurs when a conductor's temperature gradient produces a voltage across it. Thermal energy harvesting can be used in IoT devices, particularly those in industrial environments, to collect waste heat generated by machines. This environmentally friendly method helps power sensors and gadgets in places where conventional power sources might not be practicable.

11.2.3 WIND ENERGY

Wind energy is another source that is not dependent only on sunshine; much like solar energy is used to generate useful electricity. Wind energy may be used day or night, even in inclement weather like rain or clouds, unlike solar energy. Consistently windy areas, such as those close to tall buildings and bridges, offer prospects for wind energy collecting to power wireless sensor nodes used a wind turbine generator (WTG)-powered wireless sensor node [12]. With a wind speed of 3.0 m/s, the WTG produced a maximum output power of 27.3 mW, demonstrating its effectiveness and providing enough electricity to support the wireless sensor node. A WTG-powered wireless sensor node (Self-autonomous wireless sensor nodes with wind energy harvesting for remote sensing of wind-driven wildfire spread.) showed how effective it is by producing a maximum output power of 27.3 mW at a wind speed of 3.0 m/s, which was more than enough electricity to operate the wireless sensor node. Maximum Power Point Tracking was implemented in a small wind energy harvester with wireless sensor nodes to monitor wildfires, demonstrating the adaptability of wind energy harvesting in a variety of applications.

11.2.4 KINETIC ENERGY

Energy from motion is captured by kinetic energy harvesting. When mechanical stress is applied, piezoelectric materials produce voltage, whereas electromagnetic mechanisms use the electromagnetic induction principle to transform kinetic energy into electrical energy. Kinetic energy harvesting is being used in the real world in systems that capture motion from human movement or vibrations in industrial environments. For example, piezoelectric shoe inserts transform walking energy into electrical power for wearable technology.

11.3 ENERGY HARVESTER FOR IoT DEVICES

By solving the drawbacks of conventional battery-dependent devices, battery-free IoT devices constitute a substantial innovation in the field. For long-term operation, this paradigm shift depends on RF energy harvesting. IoT devices that run entirely on batteries are designed to run off standard power sources, which reduce environmental impact and does away with the need for regular battery replacements. These gadgets use alternative power sources, like energy harvesting technology, to guarantee long-term and sustainable operation.

11.3.1 RADIO FREQUENCY ENERGY HARVESTING

The process of obtaining ambient RF signals from the surroundings and turning them into electrical power to power IoT devices is known as RF energy harvesting. This technique makes use of the ubiquitous nature of RF waves, which are found in cellular, Wi-Fi, and other wireless networks shown in Figure 11.2. The advent of battery-free IoT devices, particularly those employing RF energy harvesting, represents a promising frontier in sustainable and efficient IoT deployments. While challenges exist, ongoing advancements in technology are expected to enhance the performance and viability of battery-free designs, fostering a more eco-friendly and resilient IoT ecosystem.

The principle behind battery-free IoT devices is to use alternative power sources, usually obtained through energy harvesting technologies, in order to allow long-term, independent functioning without the use of conventional batteries. A popular way to achieve battery-free operation is by using RF energy harvesting. An outline of the guiding concepts is provided below:

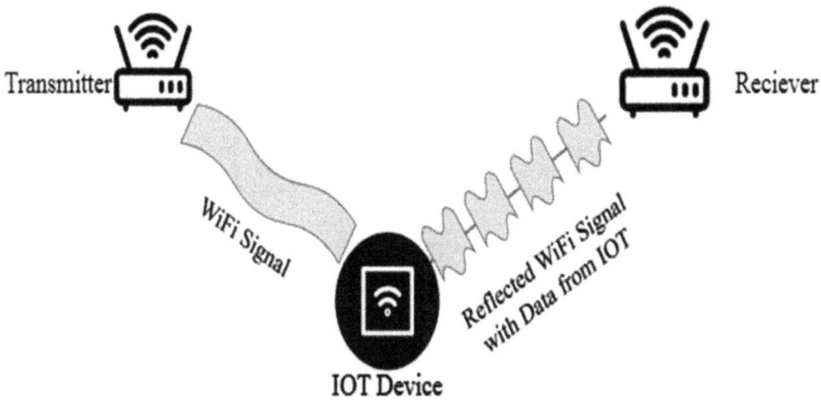

FIGURE 11.2 Power-enabled IoT device.

Source: Adapted from Ref. [13].

11.3.2 RF ENERGY HARVESTING

IoT devices that run on batteries come with antennas that are made to pick up RF signals from the surrounding air. Numerous devices, such as Wi-Fi routers, cellular towers, and other wireless communication networks, may be the source of these signals. A diode is used to rectify the acquired RF signals, which are in the form of alternating current (AC). The device has a steady power supply thanks to the diode's conversion of AC impulses into direct current (DC). Capacitors or supercapacitors are used to store the DC energy that has been captured. As reservoirs, these energy storage parts gather and store the transformed energy to offer a steady power source for sporadic usage.

11.3.3 ENERGY STORAGE

Supercapacitors, which have a larger energy storage capacity than conventional capacitors, are commonly used in battery-free IoT devices. Supercapacitors have the ability to temporarily store energy and release it in spurts as needed. Figure 11.3 shows energy moves from storage to the device's working components, intelligent energy management circuits control the flow of energy. These circuits ensure effective power distribution while optimizing energy utilization.

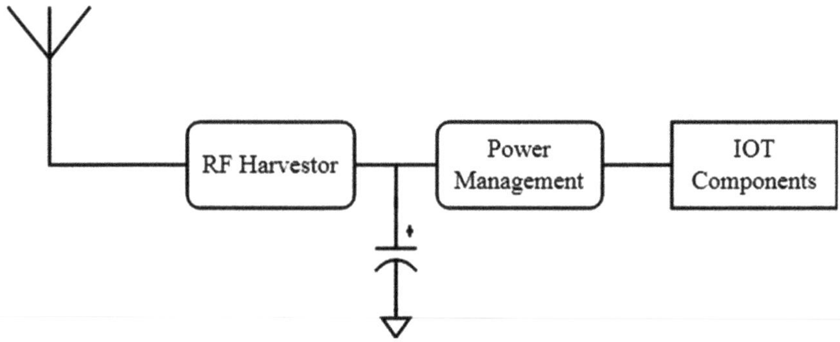

FIGURE 11.3 Block diagram of RF energy harvester.
Source: Adapted from Ref. [14].

11.3.4 OPERATIONAL MODE

Battery-free IoT devices may go into a low-power or sleep mode when not in use to save energy. In order to increase the intervals between energy collecting cycles, the gadget reduces power consumption in this mode. The gadget comes out of sleep mode to execute an instruction, gather data, or send information [14]. The IoT devices utilize their internal energy reserves to carry out these specific tasks before entering a sleep mode.

The primary goal is to estimate the harvested power from an RF signal and understand the efficiency of energy transformation [15–18]

11.3.4.1 RECEIVED POWER ($P_R X$)

The power received by an antenna from an RF signal can be calculated using the Friis transmission equation:

$$P_r x = P_t x\left(\frac{G_r G_t \lambda^2}{(4\pi)^2 d^2}\right)$$

where $P_t x$ is transmitted power, $P_r x$ is received power, G_t is gain of transmitting antenna, G_r is gain of receiving antenna, d is distance between transmitting and receiving antenna and λ is wavelength of RF signal.

11.3.4.2 RF-TO-DC CONVERSION EFFICIENCY (H)

The efficiency of converting the received RF power to DC power depends on the rectification circuit. It can be expressed as

$$Efficiency = \frac{P_{dc}}{\mathrm{P_r}\, x}$$

where P_{dc} is DC output power and $P_r x$ is received RF power

11.3.4.3 RECTENNA EFFICIENCY (H_RECTENNA)

The rectenna (rectifying antenna) is a crucial component in RF energy harvesting. Its efficiency can be calculated as

$$\eta_{\text{rectenna}} = \eta_{\text{rectification}} \eta_{\text{antenna}}$$

where $\eta_{\text{rectification}}$ is the rectification efficiency and η_{antenna} is the antenna efficiency.

11.3.4.4 ENERGY HARVESTED PER UNIT AREA (H)

The total harvested energy per unit area over time can be calculated by integrating the harvested power over time

$$H = \int_{t1}^{t2} P_{dc}(t)dt$$

where t_1 and t_2 are the begin and end time of harvesting, respectively.

11.4 CHALLENGES AND SOLUTIONS OF SELF-POWERED IoT DEVICES

Self-powered IoT devices can achieve higher efficiency, dependability, and sustainability by tackling these issues through technological advancements and all-encompassing tactics. This will help to promote the wider adoption of IoT solutions that are environmentally friendly and energy-efficient.

1. *Challenge:* Insufficient energy storage capacity makes continuous operation difficult, particularly in low-light or energy-constrained situations.
   Solution: New developments in energy storage technologies, like energy-dense batteries and high-capacity supercapacitors, increase the store capacity for dependable and extended device operation.
2. *Challenge:* Longevity and energy density of conventional batteries may be constrained.
   Solution: Energy-dense batteries and supercapacitors is bringing forth advancements that enhance the overall performance, cycle life, and energy storage of self-powered IoT devices.
3. *Challenge:* Reducing the need for frequent recharging and increasing the lifespan of devices require energy-efficient functioning.
   Solution: Self-powered IoT devices can achieve higher energy efficiency by implementing low-power components, effective energy collecting systems, and optimized communication protocols.
4. *Challenge:* Excessive power usage and ineffective power management during data processing and transmission.
   Solution: To improve power consumption and ensure efficient energy utilization, advanced power management techniques such as duty cycling and sleep modes during idle periods should be implemented.
5. *Challenge:* Ensuring self-powered IoT devices are reliable and scalable in large-scale deployments presents a challenge.
   Solution: Reliability is increased by robust communication protocols, decentralized architectures, and redundancy mechanisms. Scalability is enabled by effective energy harvesting techniques, which provide extensive deployment without sacrificing performance.
6. *Challenge:* Power from sporadic energy sources, such solar or kinetic energy, might only last some of the time.
   Solution: The total reliability of the device can be increased by implementing energy buffering methods, effective energy storage, and adaptive algorithms that can handle intermittent energy inputs with grace.
7. *Challenge:* Obtaining and transforming energy from ambient sources of different intensities is a challenge.
   Solution: Energy harvesting efficiency can be increased by utilizing multi-source energy harvesting systems and adaptive algorithms that can dynamically adapt to changing environmental conditions.

8. *Challenge:* Ensuring that self-powered IoT devices have a sustainable lifecycle that includes appropriate recycling and disposal is a challenge.
   Solution: Using environmentally friendly materials, energy-saving production techniques, and adhering to recycling guidelines all help to make these gadgets' lifetimes more sustainable.
9. *Challenge:* Taking care of security flaws in self-sufficient IoT devices, especially in low-power communication protocols, is a challenge.
   Solution: To guard against potential cyber threats and guarantee data integrity, incorporate strong security measures, encryption techniques, and safe updating systems.
10. *Challenge:* Complying with ever-changing requirements concerning environmental impact, waste management, and energy efficiency presents a challenge.
    Solution: Ensuring that self-powered IoT devices adhere to pertinent rules and fulfill industry standards will help them to meet sustainability goals for the environment.

11.5 APPLICATIONS OF SELF-POWERED IoT DEVICES

Self-powered IoT devices have applications in many different industries and provide energy-efficient ways to monitor, control, and optimize systems and processes. By lowering their need on conventional power sources and decreasing their negative effects on the environment, these devices support sustainability initiatives.

1. *Soil monitoring:* To maximize irrigation and boost crop productivity, self-powered IoT devices with sensors may track the temperature, moisture content, and nutrient levels of the soil in farming areas.
2. *Industrial health monitoring:* In industrial settings, battery-free sensors can be used to track the condition and functionality of machinery, giving real-time information for preventative maintenance.
3. *Environmental surveillance (monitoring of air and water quality)*: Self-powered IoT devices can be used to continuously monitor the quality of the air and water, allowing for the prompt detection of pollution and medical care.

4. *Wearable health trackers:* Without requiring regular battery replacements, wearable IoT devices may continuously track health by monitoring vital indications like heart rate and temperature.
5. *Smart street lighting:* Energy-saving adaptive street lighting systems are made possible by self-powered sensors in street-lights that can track ambient light, traffic, and meteorological conditions.
6. *Supply chain monitoring:* Real-time tracking of items and effective inventory management are made possible by the integration of battery-free IoT devices into logistics and supply chain management.
7. *Energy management:* In smart homes, self-powered sensors and actuators can regulate appliances, keep an eye on energy usage, and optimize it for maximum efficiency.
8. *Animal tracking:* Without causing any disturbance to natural ecosystems, battery-free IoT devices can be affixed to wildlife to follow migration patterns, analyze behavior, and support conservation efforts.
9. *Inventory management:* In retail settings, self-powered RFID tags can be used to track inventory, increasing accuracy and decreasing the requirement for human stock inspections.
10. *Smart traffic management:* To increase the efficiency of transportation, sensors installed in roads and traffic lights that do not require batteries can monitor traffic flow, identify congestion, and adjust traffic signals.
11. *Automating buildings:* To monitor occupancy, modify lighting and climate management, and improve energy efficiency, self-powered sensors can be installed in smart buildings.
12. *Handling of water:* In water distribution systems, battery-free sensors can identify abnormalities and leaks, allowing for prompt action and a reduction in water waste.
13. *Meteorological stations:* To gather data on temperature, humidity, and atmospheric conditions in real time, self-powered weather sensors can be placed in remote locations.
14. *Geological monitoring:* Remote sensing in geological applications, such as tracking seismic activity or evaluating environmental conditions in inaccessible regions, can be accomplished with battery-free sensors.

11.6 CONCLUSION

The emergence of self-sustaining IoT devices heralds a paradigm shift in the field of linked systems. Technologies like as energy harvesting, flexible electronics, and materials, along with advances in edge computing and wireless communication, have made it possible for devices to function independently, sustainably, and with ease in the future. The potential innovations that are being proposed, which include quantum energy harvesting and triboelectric nanogenerators, highlight how active this field of study is. Self-powered IoT devices have uses that go beyond traditional bounds and penetrate a variety of industries, including smart cities, healthcare, and environmental monitoring. These gadgets present a viable answer to the drawbacks of conventional battery-dependent systems by offering a more effective and ecologically friendly substitute. The integration of blockchain technology, biodegradable materials, and machine learning enhances the potential and durability of autonomous IoT ecosystems.

Future prospects are promising due to the scalability, dependability, and integration of these gadgets into daily life. The possibility of developing self-healing, adaptive, and resource-efficient systems opens doors for an IoT environment that is more responsive and resilient. Self-powered IoT devices have the potential to completely alter connectivity, accessibility, and sustainability in the digital age as the technology environment continues to change. In a future dominated by self-powered IoT devices, researchers, engineers, and innovators will collectively shape a smarter, more interconnected, and sustainable world.

KEYWORDS

- **Internet of Things (IoT)**
- **connectivity**
- **sensor technology**
- **data analytics**
- **digital transformation**
- **wearables**
- **energy access, renewable energy**

REFERENCES

1. Li, C., Wang, J., Wang, S., & Zhang, Y. A review of IoT applications in healthcare. *Neurocomputing* **2023,** *565,* 127017.
2. Memon, M. L., Saxena, N., Roy, A., & Shin, D. R. Backscatter communications: Inception of the battery-free era—A comprehensive survey. *Electronics* **2019**, *8*(2), 129.
3. Ma, D., Lan, G., Hassan, M., Hu, W., & Das, S. K. Sensing, computing, and communications for energy harvesting IoTs: A survey. *IEEE Communications Surveys & Tutorials* **2019** *22*(2), 1222–1250.
4. Zheng, Q., Tang, Q., Wang, Z. L., & Li, Z. Self-powered cardiovascular electronic devices and systems. *Nature Reviews Cardiology* **2021**, *18*(1), 7–21.
5. Tcho, I.-W., Kim, W.-G., & Choi, Y.-K. A self-powered character recognition device based on a triboelectric nanogenerator. *Nano Energy* **2020**, *70*, 104534.
6. Tsai, M.-S., et al. Self-powered, self-healed, and shape-adaptive ultraviolet photodetectors. *ACS Applied Materials & Interfaces* **2020**, *12*(8), 9755–9765.
7. Elahi, H., Munir, K., Eugeni, M., Atek, S., & Gaudenzi, P. Energy harvesting towards self-powered IoT devices. *Energies* **2020,** *13*(21), 5528.
8. Yildiz, F. Potential ambient energy-harvesting sources and techniques. *Journal of Technology Studies* **2009**, *35*(1), 40–48.
9. Almalki, F. A. et al. Green IoT for eco-friendly and sustainable smart cities: Future directions and opportunities. *Mobile Networks and Applications* **2023**, *28*(1), 178–202.
10. Sudevalayam, S., & Kulkarni, P. Energy harvesting sensor nodes: Survey and implications. *IEEE Communications Surveys & Tutorials* **2010**, *13*(3), 443–461.
11. Atallah, R., Khabbaz, M., & Assi, C. Energy harvesting in vehicular networks: A contemporary survey. *IEEE Wireless Communications* **2016**, *23*(2), 70–77.
12. Park, J.-W., Jung, H.-J., Jo, H., Jang, S., & Spencer Jr, B. F. Feasibility study of wind generator for smart wireless sensor node in cable-stayed bridge. In *Sensors and Smart Structures Technologies for Civil, Mechanical, and Aerospace Systems,* SPIE, **2010**, *7647*, 1241–1248.
13. Baali, H., Djelouat, H., Amira, A., & Bensaali, F. Empowering technology enabled care using IoT and smart devices: A review. *IEEE Sensors Journal* **2017**, *18*(5), 1790–1809.
14. Clerckx, B., Zhang, R., Schober, R., Ng, D. W. K., Kim, D. I., & Poor, H. V. Fundamentals of wireless information and power transfer: From RF energy harvester models to signal and system designs. *IEEE Journal on Selected Areas in Communications* **2018**, *37*(1), 4–33.
15. Tripathi, S. L., Ganguli, S., Kumar, A., & Magradze, T., Eds. *Intelligent Green Technologies for Sustainable Smart Cities*. John Wiley & Sons, **2022**.
16. Jadhav, K. M., & Kumar, A. Nanosecond latency drum kit. *Science, Engineering and Health Studies* **2022**, *16*, 22040005–22040005.
17. Kumar, A. Clock signal and its attribute for agriculture. *Smart Agriculture: Emerging Pedagogies of Deep Learning, Machine Learning and Internet of Things* **2021**, 198.
18. Kumar, A., Sindhwani, M., & Sachdeva, S. An innovative architecture of DRAM PUF. *Journal of Integrated Circuits and Systems* **2023**, *18*(2), 1–9.

CHAPTER 12

The Role of Internet of Things (IoT) in Hybrid and Renewable Energy Systems

SUJIT KUMBHAR[1,2,*] and SANJAY KHOT[3]

[1]*Department of Technology, Shivaji University, Kolhapur, Maharashtra, India*

[2]*Department of Automation & Robotics, Sharad Institute of Technology College of Engineering, Yadrav, Ichalkaranji, Kolhapur, Maharashtra, India*

[3]*Sharad Institute of Technology, College of Engineering, Yadrav, Ichalkaranji, Kolhapur, Maharashtra, India*

*Corresponding author

ABSTRACT

The Internet of Things (IoT) is revolutionizing the hybrid and renewable energy industry by making it possible for automated, highly intelligent and efficient electrical power generation, distribution, and consumption systems. Renewable energy systems like solar panels, wind turbines, and hydropower plants can be monitored in real time with the help of IoT devices like sensors and smart meters. Data about environmental factors (such as temperature, wind speed, and sunshine intensity), energy production levels, equipment performance metrics, and other factors are gathered by these devices. This information aids in maximizing energy production and locating inefficiencies or maintenance requirements. Systems with IoT capabilities can evaluate gathered data to forecast equipment breakdowns or maintenance needs. For instance, tracking wind turbine vibration to identify possible issues, evaluating solar panel photovoltaic (PV) cell

IoT Potential for Green Energy Solutions. Abhishek Kumar, Ajay Roy, & Shamimul Qamar (Eds.)

DOI: 10.1201/9781779643599-12

performance in order to detect deteriorating panels. IoT provides more intelligent integration of hybrid energy sources (such as solar, wind, and regular grid electricity) into a unified energy system. Key characteristics include dynamic load balancing based on real-time demand and energy availability, automated switching between renewable and nonrenewable sources to ensure energy supply reliability, as well as advanced battery energy storage management. IoT systems assist maximize renewable energy use and reduce reliance on fossil fuels, resulting in a lower carbon footprint. Efficient monitoring and control reduce energy waste, improving the overall sustainability of hybrid energy systems.

12.1 INTRODUCTION

Renewable energy sources have become essential for improving the environment in the industrial and transportation sectors. Renewable energy sources have gained widespread popularity in various countries in recent years, aided by contemporary technology, a regulatory framework, and supporting policies. The use of energy from renewable sources has increased in popularity as the worldwide demand for energy has increased [3]. There are some studies on the contribution of renewable energy to sustainable development, the implications of renewable energy on environmental security and GDP growth, and political as well as social perspectives on renewable energy, and climate change [21]. The objectives, the strategy, vision, as well as the development orientation of renewable energy in the 2030 United Nations Sustainable Development Goals (SDG) are to support everyone having recent technology, clean cooking fuels, and electrical energy, significantly increasing the consumption of renewable energy, improving progress on energy efficiency, and increasing international collaboration in services related to clean and renewable energy. The SDG-7 investigation, undertaken by five partnering organizations, namely, United Nations, World Bank, World Health Organization, International Energy Agency, and International Renewable Energy Agency, demonstrates that the world must continue to develop, excavate, and use inexpensive, dependable, eco-friendly, and efficient energy. The yearly tracking SDG-7 is derived from the energy status report 2021, with the goal of SDG-7 on energy, providing assistance to drive international policy-making and cooperation to achieve global access to sustainable energy by 2030 [17, 21].

For sustainable development, renewable energy is crucial. Nonetheless, fossil fuels still comprise the largest portion of primary energy use. In addition, there has been a consistent increase in the adoption and application of renewable energy on a global scale. By 2021, renewable energy sources contributed to 0.95 Gtoe of the world's primary energy consumption, a significant rise from 0.24 Gtoe in 2012. It was projected that renewable energy accounted for 7.4% of worldwide electricity production in 2016; by 2021, this figure had climbed to 12.8% [7].

The following energy sources were investigated: wind, photovoltaics (PV), geothermal, concentrated solar power, and biopower. Value analysis reveals that by the end of 2021, the global renewable capacity—which stood at 480 GW at the end of 2012—had increased to 1945 GW. Specifically, PV capacity has increased by 842% over the last 10 years to reach 942 GW [23].

The information analyzed from Figures 12.1 and 12. 2 demonstrates that interest in and use of renewable energy are consistently rising. Thus, it amplifies the benefits of renewable energy's contribution to climate change mitigation, environmental security, and economic growth, as well as renewable energy and sustainable development [23].

The term "Internet of Things" (IoT) pertains to the rapidly expanding category of smart, Internet-enabled devices that have connected our real-world and virtual worlds. The IoT is a system network that is built upon the Internet and being extended in order to facilitate real-time contact between machines, humans, and other objects using a variety of cutting-edge technical methods [2].

IoT principles and related innovations are now well established and understood. The emphasis must now move from proof of concept to proving proof of value—either reductions in expenses or revenue growth. For organizations to derive value from IoT in 2020, business and IT leaders must consider it as one of the several instruments in a toolbox and learn how to utilize it in concert with other equally important tools, such as analytics. According to forecasts, electronic devices connected to the Internet could account for as much as 3.5% of global energy consumption by 2027. Despite this, the IoT has the potential to help businesses improve their energy efficiency. Schneider Electric is one such example. IoT devices frequently forward data to a cloud server, during which an algorithm evaluates it and initiates an action [4, 24] (Figure 12.3).

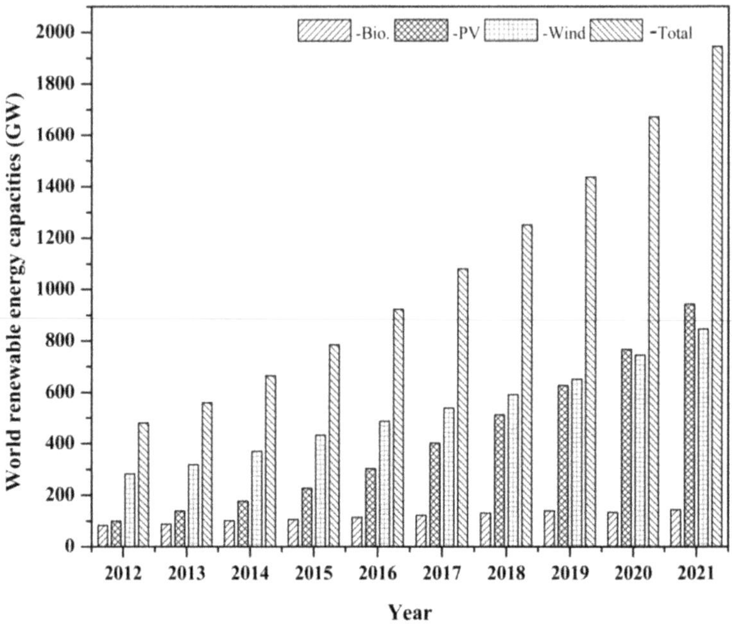

FIGURE 12.1 World renewable energy biofuels, PV, wind energy, and total renewable energy capacities.

Source: Adapted from Ref. [23].

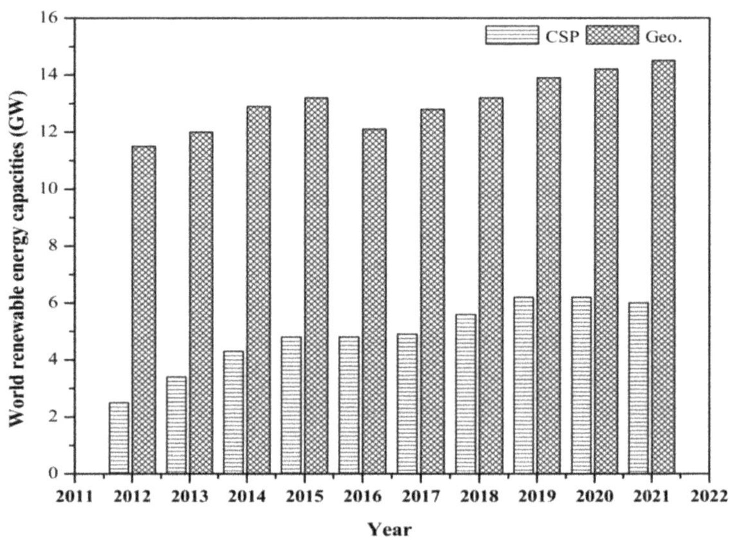

FIGURE 12.2 World renewable energy concentrated solar power and geothermal capacities.

Source: Adapted from Ref. [23].

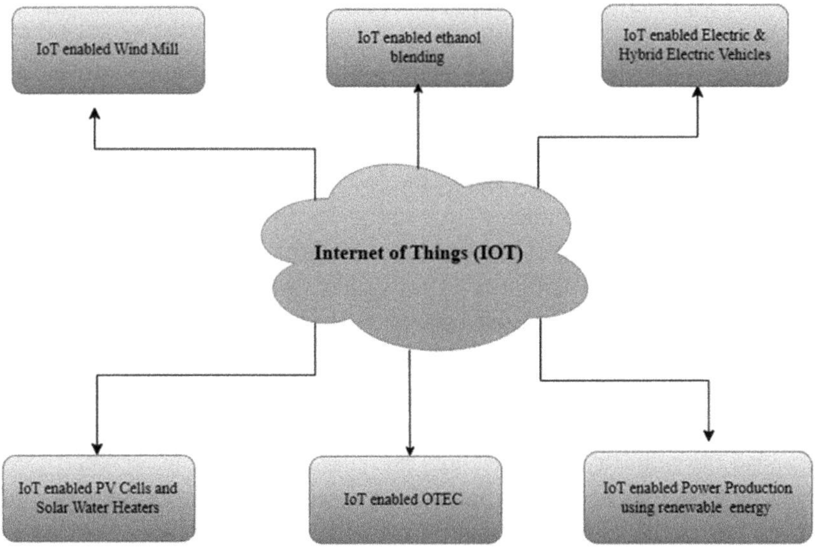

FIGURE 12.3 Role of IoT in renewable and hybrid energy systems.

12.2 INTERFACING COMPONENT

12.2.1 IOT-ENABLED SOLAR PHOTOVOLTAIC ARRAYS

Because of increased globalization as well as industrial our society now requires more energy for day-to-day tasks. The most sustainable solution to minimize stress as well as dependency on conventional fossil fuels is to harness the energy from the sun. Solar energy has been distinguished by its low cost, environmental friendliness, absence of regular maintenance and, most significantly, free maintenance. Nevertheless, monitoring solar installations is required to enhance output power by establishing angles with the sun's position in real time. This is simply accomplished through the use of IoT technologies. The IoT is often used to measure the efficiency of solar energy. It is also used to keep the solar plant functioning. Solar energy installations need to be monitored to ensure efficient electricity output. This aids in the restoration of economic power production from power plants by repairing defective star panels, searching for contacts, and alternating those difficulties with sludge accumulated on output decreasing panels and shaking wire [10]. On the other hand, a machine-controlled IoT, which is essentially a different kind of energy monitoring gadget, allows for automated use of alternative energy sources through the Internet from any place. The new technology constantly assesses the solar battery and immediately sends the capacity performance of the IoT system *via* the Internet. We have a tendency of employing IoT thing communicate to send alternate energy data to the net of the IoT thing speak server. The design now displays these parameters in a user-friendly graphical user interface and notifies the user when the performance limit is reached. This enables convenient remote surveillance of solar plants while also assuring optimal power generation. The global cost of renewable energy technologies is reducing, which supports the construction of large-scale solar plants. Complex technologies reliant on Internet connectivity are required for the automation of plant monitoring on such a large scale of initial installations because the majority of field units are situated in remote, isolated locations and are thus not coordinated from a central office. Through the use of IoT, the project is predicated on the utilization of the most advanced, economical method for remotely monitoring a solar plant's performance. It can support real-time monitoring, troubleshooting, and plant maintenance [9].

A PV system is made up of solar modules, power converters, and storage devices. It is nothing more than an energy harvester that converts sunlight into electricity. The conventional strategy of producing power using fossil fuels is very different from this method of producing energy utilizing PV systems. On the other hand, energy distribution and transmission employ similar procedures. PV generators are formed by connecting arrays of PV modules to form PV arrays, which are placed in series and parallel configurations. These PV generators are built in such a way that they immediately start producing electricity as soon as exposed to sunlight. These PV generators produce DC electricity, which power converters subsequently transform into alternating current electricity. Depending on the needs, the energy can subsequently be transmitted *via* a transmission network to the electrical grid or self-consumed by a particular demand. Even if energy is produced and transmitted, battery storage may be necessary in some situations. PV systems are grouped into two groups based on their components and operation: off grid PV systems and on grid PV systems. Figure 12.4 depicts a schematic picture of the operation of a solar system with IoT technique [13].

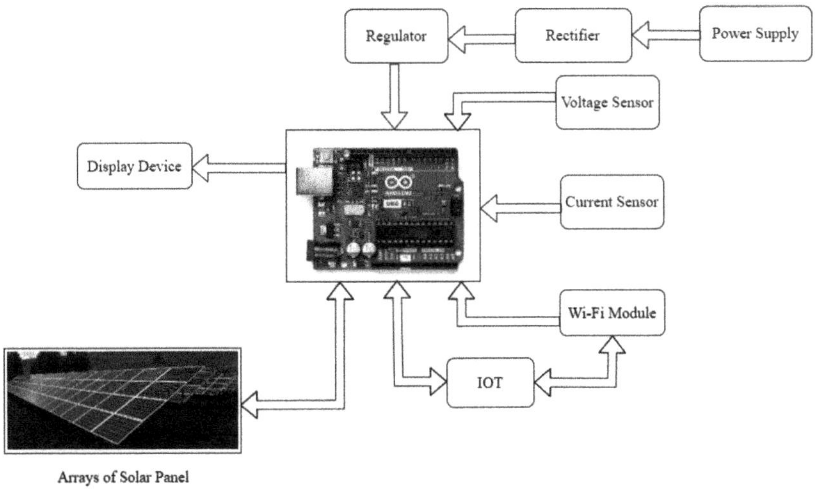

FIGURE 12.4 IoT-enabled solar photovoltaic arrays.

IoT is a technology that was created by combining "Internet, micro-electromechanical systems, wireless technologies, and micro services." The interconnection of similar objects with unique IDs, including digital, mechanical, and computer components, is the essence of the IoT. This coordination aids in data transmission throughout the entire network by bridging the separation of information technology as well as operational technologies, eliminating the need for human-to-human and computer-to-human communication. IoT can tackle the most complicated problems that modern scientific and engineering systems encounter. There exist various components in solar power systems that exhibit varying operational behaviors. To be more precise, the system cannot produce power continuously because solar energy fluctuates with time and is weather-dependent. This has an indirect impact on the operation of other components of the system such as power converter voltage levels, battery life, electrical energy requirement as per load variation, and so on. Dust deposition and different environmental considerations occasionally cause the solar system to perform poorly. However, in the long run, these issues lead to the system failing. Employees find it difficult to monitor such failures since they need to frequently visit the plant site and gather operational data. It is a time-consuming task when the plant is located in a remote location.

As a result, fixing these issues takes a long time and is sometimes not possible owing to a lack of relevant information regarding the system malfunction or poor operation. Therefore, in order to track system parameters and save data on a cloud platform, a system with continuous monitoring needs to be deployed in conjunction with the PV system. Understanding its performance and the reasons behind poor performance would be made easier with the help of the data saved. This makes it possible to do diagnostic and maintenance procedures when performance is poor due to issues. As a result, the requirement for IoT is critical in performance monitoring and adjusting system settings via remote control.

PV systems' IoT architecture consists of three layers: the gateway linkage, the remote monitoring and control layer, and the PV system design environment. The PV system design environment, comprising all components coupled in suitable configurations to satisfy user requirements, is the initial layer. In this scenario, the next layer of the IoT architecture, the Arduino server, is connected to the PV system components. The second layer, known as gateway linkage, employs a wireless router equipped with Internet firewall functionality to connect the web server and PV hardware configurations. The primary element that facilitates the implementation of a web server *via* an Ethernet connection or wireless networking module is the Arduino server. The Arduino server's microcontroller monitors, controls, and maintains an eye on the performance of the hardware components of the PV system. The third layer, or remote monitoring and control layer, will receive the data from the server. The PV system's acquired data will be sent by the server to storage devices, allowing for the periodic generation of reports. Users can use the Android interface using cloud data *via* Wi-Fi network to get the information in the form of visual graphs or summaries [18].

12.2.1.1 IoT-ENABLED ETHANOL BLENDING IN GASOLINE

Bioethanol is a renewable fuel made from fermenting crops. It can be combined with gasoline or used on its own to power vehicles. South Africa farms a lot of sugarcane, which could be used to make bioethanol, but the costs have kept the country from producing it on a significant scale. Higginson wanted to use artificial intelligence (AI) to optimize production in a 100-L steel fermentation tank. The amount of nourishment provided to the yeast at the start of a batch is critical to the fermentation process.

If the sugar originates from a specific batch of cane or molasses, the nitrogen content may be more or lower than predicted, making prediction and management problematic. Higginson developed neural networks to anticipate the end ethanol concentration based on the early stages of fermentation, and then used that information to add additional sugar or dilute the existing sugar throughout the process to boost the ethanol concentration. Higginson devised a method for remote experiment observation. Sensors that recorded characteristics like temperature and density were linked to two Arduino microcontrollers and a programmable logic controller (PLC). A Raspberry Pi, a different little device, served as a server, gathering data from the Arduino boards and transferring it to the PLC, the system's primary controller. Typically, he would keep an eye on the procedure using supervisory control and data acquisition (SCADA) software on a lab computer, but that machine was not connected to the Internet. He used ThingSpeak™, an IoT analytics platform that gathers data and visualizes it graphically, which is accessible to all university students *via* their Campus-Wide License for MATLAB, to receive data from the Raspberry Pi in order to remotely monitor the experiment. Users of ThingSpeak can view the data in real time on any web-enabled device. Higginson imported the data from ThingSpeak into MATLAB for additional analysis once the experiment was over. He used to have to copy a spreadsheet from the SCADA PC to a flash drive in order to obtain data from the experiment [6].

12.2.2 IoT-ENABLED ELECTRIC AND HYBRID ELECTRIC VEHICLES

Worldwide, the number of electric cars (EVs) is rising daily. Electric vehicles (EVs) are getting more and more popular because they emit no CO_2 and are eco-friendly. One of the primary problems that needs to be fixed in its entirety is energy storage and charging systems. the use of AI and the IoT to fully autonomously drive EVs by utilizing a variety of networked sensors and monitor charging system performance. It is possible to build a self-charging system and have the car communicate with its surroundings. AI is the ability of the human mind to carry out activities and make decisions using various forms of logic and programming, just like human intelligence. Automation of EVs is accelerated by AI. Future fully autonomous driving cars powered by IoT and AI will allow us to transform cities into smart ones and lessen traffic, parking, and battery charging problems [15].

EV manufacturers have adopted new technologies such as AI and the IoT to increase productivity, develop vehicle manufacturing, and transition to a smart car environment. Compared to conventional vehicles, EVs produce fewer greenhouse gases and air pollutants during their lifetime. We are aware that there is a finite amount of fuel on Earth, so we must find another source. The best source of electricity comes from EVs. To fully use and outperform traditional vehicles that have detrimental impacts on the environment, hybrid electric vehicles (HEVs) are the ideal choice. Different methods are used today to charge EVs, including hybrids. Plug-in charging requires a plug to be connected to the vehicle, whereas wireless charging systems do not require a plug and can be accomplished through inductive charging. Additionally, we charged the battery using a solar charging apparatus. The capacity of a machine to carry out intelligent activities is known as AI. AI is a computer-based system that produces data containing information such as navigation, voice recognition, vehicle speed, time, motion detection, image recognition, driver monitoring, battery charge monitoring, and virtual assistance. AI uses sensors, actuators, and software to perform tasks similar to those of the human brain. IoT is primarily regarded as a methodology for managing linked devices and their data in autonomous HEVs. Therefore, using various sensors integrated within it, it is simple to monitor battery charging, traffic, parking, and weather conditions with the use of IoT. Additionally, the presence of advanced sensing technology in these EVs enables them to effectively monitor various traffic situations, making them invaluable resources during congested periods on the road. The IoT make use of common objects that have easy-to-use connectivity and human control. Researchers have examined the self-charging system in this work using a variety of charging methodologies, such as autonomous driving. It is required to apply IoT and AI-based technology because this technique takes into account a data transfer methodology where the analyzed information will be delivered to the control center [14].

Figure 12.5 describes the layout of the electric car's IoT-based battery monitoring and management system. In this prototype system, the micro-controller ESP32 has been used to implement it. The USB port on a PC is used to supply electricity or any other source that provides the system with the necessary 5 V DC. Making use of the user's smartphone's 4G or 5G connectivity as an Internet source, transfer data through the ESP32 to the server that hosts the IoT platform. Use the ESP32's integrated Wi-Fi module to connect to the Internet. An ESP32 was paired with one current sensor to measure the current flowing into the battery and another to measure the current flowing out of the battery. A fuel gauge sensor is used to determine the capacity or state of charge (SoC) of a Li-ion battery. A Li-ion battery charger is attached to the battery to initiate charging, and a relay controls charge operation. As illustrated in Figure 12.3, the system notifies the user through a mobile application whenever a battery's current SoC falls below 30%. This prolongs the battery's life by preventing harm from over discharge. The charging procedure ends once the battery is completely charged or when the off switch is pushed. The lithium-ion battery charger stops the charging process *via* an inbuilt protected circuit. Two current sensors, a battery fuel gauge module, and the ESP32 are all connected to the same serial port over the I2C protocol in order to gather and process data. Two bi-directional open-drain lines are all that are required for SDA and SCL data transfer. We've turned up these two lines. The clock signal is transmitted *via* a serial clock (SCL), and data is transported by serial data (SDA). The SCL line's high-to-low pulse synchronizes each data bit sent on the SDA line. ESP32 obtained the current in and out value from INA219, the battery SoC through MAX 17043, and so forth. Charger efficiency is increased by the relay that connects the battery to the TP4054.

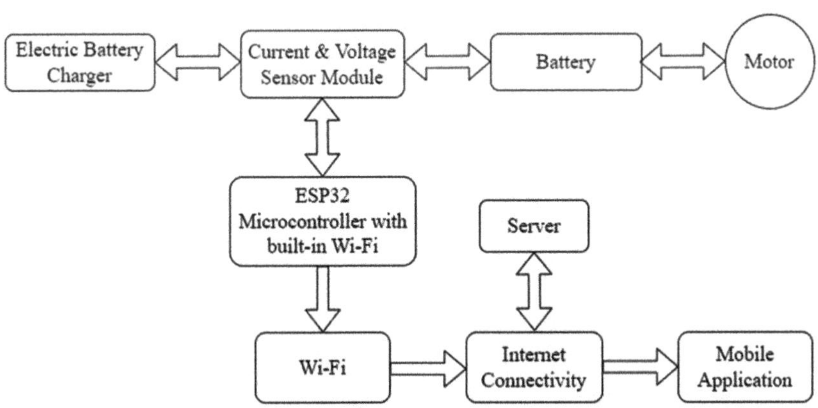

FIGURE 12.5 IoT-enabled electric and hybrid system.

12.2.3 ESP32 MICROCONTROLLER

ESP32 is composed of two main components: the 40 nm-processed Xtensa LX6 processor. CPU cores individually can be managed. There is 520 KB of on-chip SRAM accessible for instructions and data. For instance, the ESP32-Grover features 8 MB of SPI PSRAM (Pseudo Static RAM) and 4 MB of external SPI flash.

12.2.4 IoT-ENABLED SOLAR WATER HEATING SYSTEM

For all living things, including humans, water may be a resource that is vital to their way of living. Taking a bath is one of the most common daily activities using water. For maximum comfort when bathing, most people need a shower with the right temperature of water. To do this, the previously requested swimming water must be obtained by using a storage tank. But, one should anticipate spending a lot of time in the situation prepared by using a storage tank before taking a bath. One workaround is to tell them to point to the distant storage tank instead of the storage tank when it is not required for them to be in the same location. As an outcome, a degree of personality comfort will arise from having the vessel in a state that allows it to be utilized at the users' preferred temperature [20].

Modern civilization is continually evolving due to the IoT, from how individuals communicate with one another to how they go about their daily lives. IoT may consist of a vast network of interconnected devices. The IoT makes it possible to perceive an object through networking infrastructure. Bathing is one of the most prevalent human needs, and these gadgets gather and communicate information on how they are used and consequently the environment in which they are utilized. In addition to washing away grime, taking a bath can be a soothing approach to loosen up tense muscles in the body. More hot water is needed in the bathtub in order for the body to relax. Most of the time, warm water is first brought to a boil in a pot or pan. Once the water reaches a boiling point, transfer it to the tub's cold water reservoir. This approach is unsuccessful and energy intensive in its current configuration. It is unable to change the temperature to meet its requirements. There are currently a lot of water heater solutions on the market that can be bought to address the issues. The idea behind the gadgets is that they use two containers, one of which is a water heater. These gadgets do, however, have several drawbacks, such as the traditional on/off technique and the lack of a monitoring system that may save users time [16].

To gather energy from the sun, the solar water heater is equipped with an array of interconnected solar panels. The tank for holding water is housed within the collectors. The water was warmed during the day. The principle of gravity dictates that cold water gets heated in a circular fashion because hot water has a lower density than cold water. For later usage, the hot water is transferred to the tank. The basic diagram for the IoT enabled monitoring system is shown in Figure 12.6.

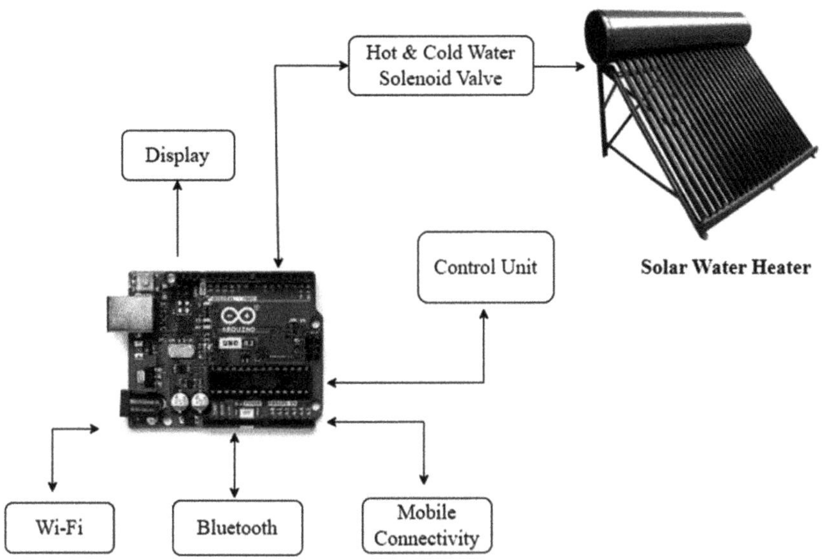

FIGURE 12.6 IoT-enabled solar water heating system.

An inspiration layer, an absorption layer, an exterior and inner tube, and a vacuum space make up a vacuum tube. Water is kept in reserve inside the inner tube. While the selective absorption layer transforms the energy into thermal energy, the cavity serves as an insulator to keep heat energy from escaping. The flat screen mount is housed in an insulated enclosure that is shielded by tempered glass. For both household and commercial solar heating applications, the flat solar collector is perfect. Even though it is one of the lightest flat panels available, it yet provides the most warmth. The flat collectors for transfers are made of sturdy metal to resist large loads and fluctuating weather conditions. The output of pipe is equipped with a temperature sensor. The tank and pipe sensors are connected by a microcontroller. There are placed solenoid valves on the storage tank and solar outlet. Those valves are connected to the temperature sensors. In response to the required temperature, these sensors will activate the solenoid valve, which, depending on the temperature disparity, will release either hot or cold water. The complete process can be managed by the IoT. A minute feature like each user's temperature changing at a different moment, the amount of time a solar heater works, or any other facility can be controlled by an app [1] (Figures 12.7 and 12.8).

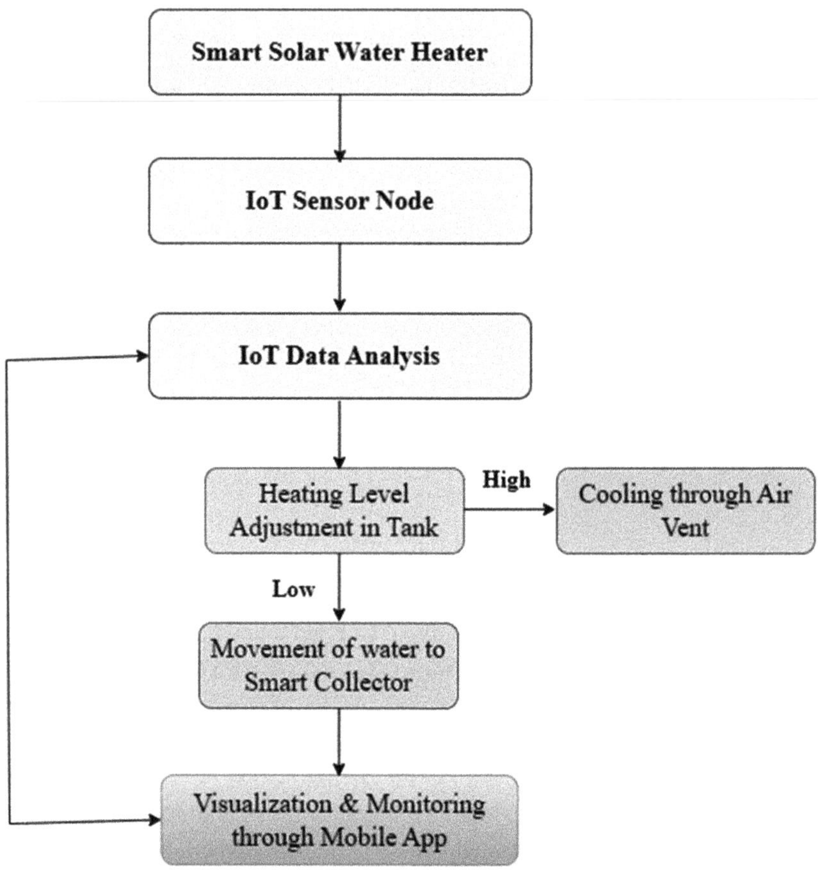

FIGURE 12.7 Flowchart of IoT-enabled solar water heating system.

When compared to the prior water heater system, the IoT-enabled solar water heater system reduces water waste. This IoT-enabled smart solar water heater keeps an eye on temperature, pH level, and tank level all the time. The data that the IoT node gathers is uploaded and broadcast to the cloud. By analyzing the sensor value, the IoT cloud provides instructions to the control unit of the smart solar system. For smart homes and Industry 4.0, the efficiency of the IoT-enabled solar water heating system is excellent. Additionally, by examining the data acquired regarding the measured data, the user can ascertain whether or not the system is operating as intended. To further develop this work, a board like IoT edge-enabled devices would be appropriate [18].

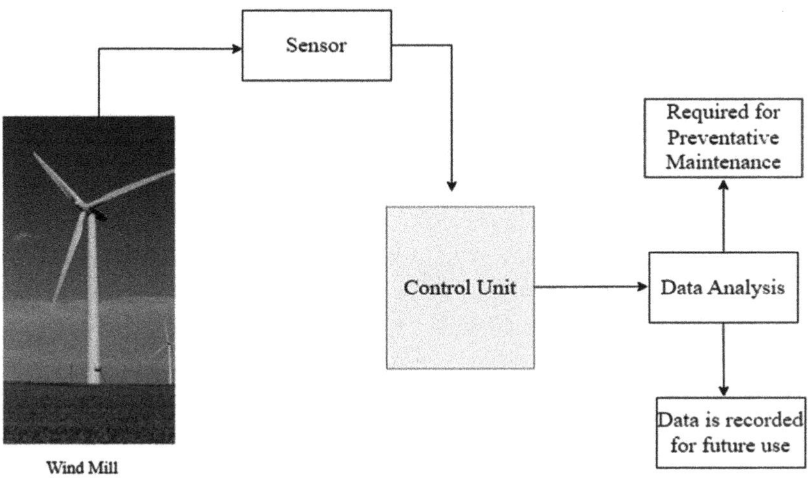

FIGURE 12.8 IoT-enabled wind turbine.

12.2.5 IoT-ENABLED WIND MILL

Wind turbines are a vital part of the renewable energy generating process. By 2030, wind farms—which are made up of numerous wind turbines—are expected to provide 15% of all electricity. Wind turbines are typically erected in isolated locations with lots of wind. Mountaintops and areas close to the coast are two examples of these locations. A wind turbine has a number of sensors built into it to measure its state for maintenance and monitoring. However, remote communication access to the wind turbines is necessary for convenient state monitoring. A network built on the IoT may be a workable answer to this issue. In order to track turbine performance and detect issues, a real-time monitoring system may be employed when wind turbines are connected to an IoT network. This would enable staff members to act promptly. Given that communications between wind turbines and the Internet will likely take place over wireless communication channels, a reliable communication infrastructure is required between the wind turbines as well as wireless base stations or access points that are connected to the Internet. In addition, a fusion algorithm must be created to accurately estimate the state of the system while accounting for the uncertainty of the wireless channel in order to process and integrate the various sensor readings. Because the control centers are located in distant areas, they are typically several hours distant from the wind farms. Control centers can monitor and regulate the performance of wind turbines with the aid of remote data transmission connectivity through an IoT network connected to the Internet. Wind turbines are located in remote areas, hence wireless networks like satellite or cellular networks are needed for IoT network access. Figure 12.6 shows an integrated scenario with a wireless IoT network and wind turbines [22].

Wind energy is employed as a backup energy source to fulfill the excess energy requirement. Effective utilization of wind energy requires regular maintenance of wind mills. Reconfiguring wind mills becomes a challenging undertaking whenever a malfunction arises. Therefore, we need a decent method to accomplish it. Certain data are needed to control the maintenance plan and make maintenance easier. It was formerly monitored. If it were possible to remotely access the data needed for appropriate maintenance, the procedure might be simplified. It also helps when the device may be programmed to switch on or off (in unfavorable conditions) based on the data gathered. It reduces a great deal of human labor (manual monitoring). A network of interconnected computing devices known as the IoT-enabled data transmission and reception across a network of devices without the need for human involvement [19].

IoT data analysis is possible at nearly real-time speed. Furthermore, certain decisions can be made in light of the data analysis. The main goal is to create an "IoT-based Windmill monitoring system" that will gather the information needed to keep an eye on a windmill and use that information to automate and calculate its maintenance plan. Several other helpful pieces of information are gathered from the data. To continually track the scattered wind power plant's efficiency and operational status, a real-time, low-cost remote monitoring system is suggested. The LoRa network serves as the foundation for this system, which tracks variables linked to the wind turbine's output performance and operational state. The long-range transmission protocol is utilized by the wind power generator monitoring node to facilitate regional data transmission and monitoring of wind power generators. A battery of tests validates the behavior and functionality of this proposed system. The radio-frequency mesh network is used by the control and monitoring system to collect data from numerous devices (turbines), and an Access Node transmits that data to a back end system. These radio-frequency mesh networks consist of radio nodes clustered around access nodes that serve as WAN gateways to a data center running back-end system software for communications, data acquisition, scheduling, and data storing, processing, presenting, and exporting. A WSN-based interactive embedded system is being developed to connect multiple windmill sensors in an efficient and dependable manner, enabling remote monitoring from any place. The CMS is made up of a number of sensors that continuously gather operational data from the windmill, such as wind speed, temperature, humidity, vibration, rotor health, and generator health. The sensors then use a wireless network to transmit the data to a the internal controller for local monitoring [13].

12.2.6 IoT-ENABLED OCEAN THERMAL ENERGY CONVERSION

Everyday things are connected to both wired and wireless networks in the IoT, which is principally an information sharing ecosystem. It has been applied in a number of industries recently, including energy systems, industrial security, smart homes, smart cars, smart cities, and healthcare. It uses are not limited to consumer devices and appliances. One important renewable energy source at the moment is ocean thermal energy. An increasingly viable option for a future sustainable energy supply in coastal areas is ocean thermal energy. For every ocean thermal power plant, remote monitoring is necessary because on-site monitoring is not feasible for the installer. Currently, utilizing digital technology to monitor Ocean Thermal Energy Conversion (OTEC) plants and leveraging IoT power. The fluctuation of power generation from ocean thermal power plants is caused by variations in temperature, sun irradiation, and many other factors. For this reason, remote monitoring is essential. The IoT strategy is used to design a remote monitoring system for the OTEC plant [17]. This method actually envisions a time not too far in the future when everyday objects will be fitted with microcontrollers and transceivers for digital communication. IoT-based systems make significant strides toward monitoring via intelligent decision making from the web, and remote monitoring eliminates the risks connected with traditional wiring systems. It also greatly simplifies and lowers the cost of the data measurement and monitoring process. The decentralized architecture and flexible deployment options of remote monitoring systems make them ideal for both home and commercial use. To retrieve, analyze, transmit, control, and offer feedback on the remote information, remote monitoring systems typically need to leverage the most recent advancements in science and technology in the disciplines of communication technology and other related fields. Additionally, it combines the extensive use of computer software, electronics, and instruments. Frequent observation of the current OTEC system approaches provides certain challenges, such as low real-time and low automaticity. An effective system for monitoring and controlling distant environment information can help prevent these issues. The OTEC station and automated diagnostic methods should be a part of this system [12].

The Ocean Thermal Power Plant's IoT application schematic is depicted in Figure 12.9. The three structured schematic diagram begins with the sensing layer at the bottom, which is made up of voltage, current, and pyranometer sensors for measuring irradiance, among other sensors. Additionally, this layer uses microcontroller-based data processing to manage the sensor data that is gathered. The microcontroller initiates and transmits data to the server through communication with the wireless module. The second layer, often known as the network layer, is where database storage and data recording from the plant are supposed to occur for real-time processing. The application layer, which is the layer that comes after the network layer, then makes use of this conserved and analyzed data. This layer designs complex web-based services using the gathered, processed, and stored data. A visual user interface will assist in keeping an eye on the plant's performance. In order to reduce the time required for decision-making, the console will also give the administrator advice on choices based on historical data. It will be simpler to monitor an ocean thermal power plant's entire performance using a web-based method by utilizing an IoT-based remote monitoring system [5].

12.2.7 IoT-ENABLED HYBRID SYSTEMS

Renewable sources of energy like wind and PV are receiving a lot of attention due to their energy capacity and technological maturity. This is because the goal is to reduce greenhouse gas emissions and increase the total availability of renewable energy sources. When it comes to supplying electricity to remote villages and independent people, these sources take the place of conventional energy sources. Systems could operate in both a standalone and a grid-connected mode by configuring various energy arrangements. Due to their intermittent nature, a significant number of renewable energy sources are challenging to integrate into the electricity system [8]. A schematic diagram for the hybrid renewable energy system's grid integration is shown in Figure 12.10.

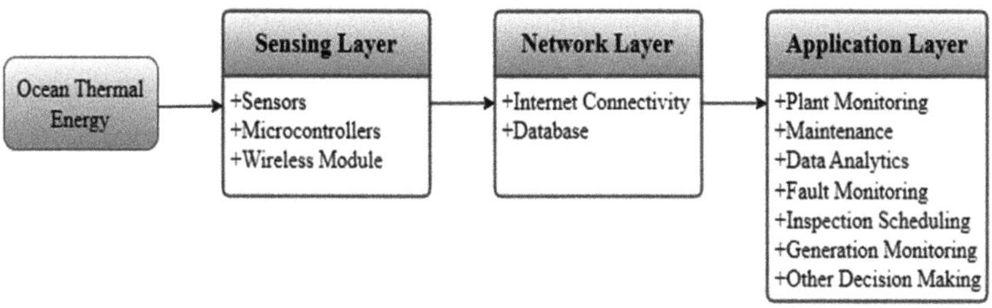

FIGURE 12.9 IoT-enabled OTEC.

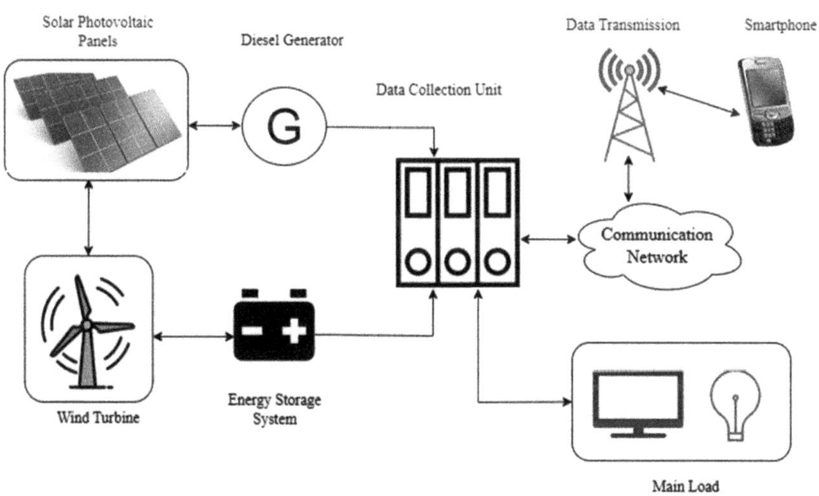

FIGURE 12.10 IoT-enabled hybrid renewable energy system.

Through this integration, the hybrid renewable energy system will support many services, including response to demand and demand-side management, and offer a multitude of benefits to utilities that provide electricity during periods of maximum demand. The integration of distributed energy resources with enhanced resilience, dependability, and efficiency will be made possible in large part by the underlying communication network. The many energy sources (such as windmills, PV cells diesel-powered generators, and rechargeable batteries), transformers, feeders, converters, and electrical lines make up the power infrastructure layer. Through an underlying communication network that connects sensor and actuator nodes, the communication infrastructures overlay support the physical infrastructure [11].

Communication infrastructures are essential components of hybrid renewable energy systems; they provide data flow between controllers, the control center, and data resources (meters and sensors). The system architecture is defined by the information flow from various entities in order to provide control and monitoring from a distance operation. Monitoring, communication, interpreting, and actuation will all be greatly facilitated by IoT technology to support a variety of microgrid applications. First, a communication network is used to gather and send measured data to the local control center. Following the decision-making process based on this data, control commands are transmitted via the communication network to controllable devices [5].

12.3 CONCLUSION

The IoT is the next evolution of the Internet, allowing connections and communication between everything at anytime and anywhere. IoT is one of the most promising and future technologies in the world, according to continuing study. There is a detailed explanation of the significance of IoT and how it will raise living standards and benefit the next generation. We have talked about the main enabling technologies that help with the real-world use of IoT. The present and potential uses of IoT across a range of industries have been emphasized, along with the ways in which these uses might be combined to create more intricate and advantageous uses. The use of IoT for solar power plant monitoring is a significant step in the ongoing integration of renewable energy sources into the utility grid. Therefore, the process of making decisions about large-scale solar power plants and their grid integration in the future will be improved by the automation and intellectual development of solar energy generation monitoring.

In commercial applications, this solar-wind energy source combination will be quite successful. It avoids lightning-related mishaps and is environmentally beneficial. It is employed to stop sudden power surges. Given the minimal maintenance requirements of this system, there is potential for a decrease in electricity expenses. Additionally, incidents of load shedding or power outages are nonexistent. Furthermore, the system is managed by the IoT, allowing the site manager to obtain comprehensive facility information on-site and facilitate effective maintenance for routine inspections and malfunctions. It can be incredibly economical and dependable.

KEYWORDS

- **IoT**
- **hybrid and renewable energy system**
- **photovoltaic panel**
- **wind energy**
- **bioenergy**
- **OTEC**

REFERENCES

1. Benhamaid, S., Bouabdallah, A., & Lakhlef, H. Recent advances in energy management for Green-IoT: An up-to-date and comprehensive survey. *Journal of Network and Computer Applications* **2022**, *198*, 103257. https://doi.org/10.1016/j.jnca.2021.103257
2. Constance, E. The Internet of Things: Preparing for the revolution. *Journal of Cyber Policy* **2017**, *2*(2), 152–154. https://doi.org/10.1080/23738871.2017.1361890
3. Elani, U. A., Alawaji, S. H., Hasnain, S. M., & Arabia, S. The role of renewable energy in energy management and conservation. *Renewable Energy* **1996**, *9*(1), 1203–1206.
4. Hu, J., van der Vlist, B., Niezen, G., Willemsen, W., Willems, D., & Feijs, L. Designing the Internet of Things for learning environmentally responsible behaviour. *Interactive Learning Environments* **2013**, *21*(2), 211–226.
5. Huang, R., Yang, X., & Ajay, P. Consensus mechanism for software-defined blockchain in internet of things. *Internet of Things and Cyber-Physical Systems* **2023**, *3*, 52–60.
6. Jain, S., Choudhari, P., & Shrivastava, A. Things : Architectures, enabling. In: *Healthcare Paradigms in the Internet of Things Ecosystem*. Elsevier Inc **2021**. https://doi.org/10.1016/B978-0-12-819664-9.00001-6

7. Khare, V., Khare, C., Nema, S., & Baredar, P. Renewable energy system paradigm change from trending technology: A review. *International Journal of Sustainable Energy* **2021**, *40*(7), 697–718. https://doi.org/10.1080/14786451.2020.1860043
8. Kumar, S., Kaur, T., Upadhyay, S., Sharma, V., & Vatsal, D. Optimal sizing of stand alone hybrid renewable energy system with load shifting. *Energy Sources, Part A: Recovery, Utilization and Environmental Effects* **2020**, 1–20. https://doi.org/10.1080/15567036.2020.1831107
9. Li, L., & Wang, X. Design and operation of hybrid renewable energy systems: Current status and future perspectives. *Current Opinion in Chemical Engineering* **2021**, *31*, 100669. https://doi.org/10.1016/j.coche.2021.100669
10. Malche, T., & Maheshwary, P. Internet of things (IoT) for building smart home system. In: *Proceeding of International Conference on I-SMAC (IoT in social, mobile, analytics and cloud)(I-SMAC)* **2017**, pp. 65–70. IEEE.
11. Mashrur, M., Das, B. K., Das, P., Hossain, S., & Kibria, G. Energy management and sizing of a stand-alone hybrid renewable energy system for community electricity, fresh water, and cooking gas demands of a remote island. *Energy Conversion and Management* **2024**, *299*, 117865. https://doi.org/10.1016/j.enconman.2023.117865
12. Mendez Mena, D., Papapanagiotou, I., & Yang, B. Internet of things: Survey on security. *Information Security Journal* **2018**, *27*(3), 162–182.
13. Mukta, M. Y., Rahman, M. A., Asyhari, A. T., & Alam Bhuiyan, M. Z. IoT for energy efficient green highway lighting systems: Challenges and issues. *Journal of Network and Computer Applications* **2020**, *158*, 102575. https://doi.org/10.1016/j.jnca.2020.102575
14. Qahtan, M. H., Mohammed, E. A., & Ali, A. J, *IoT-Based Electrical Vehicle ' s Energy Management and Monitoring System* **2022**, *9*, 1–15.
15. Rani, & Malik, A. A social relationship-based energy efficient routing scheme for opportunistic internet of things. *ICT Express* **2023**, *9*(4), 697–705.
16. Razdan, S., & Sharma, S. *Internet of Medical Things (IoMT): Overview, Emerging Technologies, and Case Studies* **2022**, https://doi.org/10.1080/02564602.2021.1927863
17. Rezaee, N., Zanjirchi, S. M., Jalilian, N., & Bamakan, S. M. H. Internet of things empowering operations management; A systematic review based on bibliometric and content analysis. *Telematics and Informatics Reports* **2023**, *11*, 100096. https://doi.org/10.1016/j.teler.2023.100096
18. Singh, N., Raza, M., Paranthaman, V. V., Awais, M., & Khalid, M. Internet of things and cloud computing. In: *Digital Health. INC.* 2023 https://doi.org/10.1016/B978-0-12-818914-6.00013-2
19. Singh, V. K., Rishav, M. B. C., & Ray, U. K. IOT based windmill monitoring system. *IJERT NCESC* **2018**, *6*, 13.
20. Taiwo, O., & A. E. Ezugwu, Internet of things-based intelligent smart home control system. *Security and Communication Networks* **2021**, *2021*(7), 1–17.
21. Trinh, V. L., & Chung, C. K. Renewable energy for SDG-7 and sustainable electrical production, integration, industrial application, and globalization: Review. *Cleaner Engineering and Technology* **2023,** *15***,** 100657. https://doi.org/10.1016/j.clet.2023.100657

22. Yaqub, U., Al-Nasser, A., & Sheltami, T. Implementation of a hybrid wind-solar desalination plant from an Internet of Things (IoT) perspective on a network simulation tool. *Applied Computing and Informatics* **2019**, *15*(1), 7–11.
23. Yolcan, O. O. World energy outlook and state of renewable energy: 10-year evaluation. *Innovation and Green Development* **2023**, *2*(4), 100070.
24. Zhang, H., Li, F., Wang, J., Wang, Z., Sanín, C., & Szczerbicki, E. Experience-oriented intelligence for internet of things. *Cybernetics and Systems* **2017**, *48*(3), 162–181.

CHAPTER 13

IoT-Enabled Energy Monitoring and Management Systems

SUJIT KUMBHAR[1,*], GOVIND SINGH PATEL[2], PRATPSINH CHOUGULE[2], and YOGESH KAMBLE[2]

[1]*Department of Technology, Shivaji University, Kolhapur, Maharashtra, India*

[2]*Department of Automation & Robotics, Sharad Institute of Technology, College of Engineering, Yadrav, Ichalkaranji, Kolhapur, Maharashtra, India*

*Corresponding author

ABSTRACT

Energy monitoring and management technologies powered by the Internet of Things (IoT) are transforming the way energy is generated, distributed, and being consumed. These systems use IoT technology to deliver real-time data, automate operations, and optimize energy consumption in the residential, commercial, and industrial sectors. IoT energy consumption and generation are monitored in real time by sensors and devices, which track characteristics such as voltage, current, temperature, and equipment performance. IoT gateways collect data from sensors and send it to centralized systems or cloud platforms. Cloud and edge computing store and process a vast amount of data for actionable insights while also providing scalability and low-latency decision-making. Control and automation systems allow for automated responses, such as adjusting loads or switching energy sources, to increase efficiency. Energy monitoring and management solutions powered by the IoT are critical to a sustainable energy future. These systems help optimize energy use, integrate renewables, and promote

IoT Potential for Green Energy Solutions. Abhishek Kumar, Ajay Roy, & Shamimul Qamar (Eds.)

DOI: 10.1201/9781779643599-13

optimal energy consumption across several industries by providing real-time visibility, predictive capabilities, and automation.

13.1 INTRODUCTION

The number of connectors and other linked devices is rising along with the usage of the Internet and its range of applications. These days, all kinds of organizations—academic, research, business, personal, etc.—use the Internet. Further research in related topics is motivated by the Internet's rapid development and advancement [1]. The contemporary generation has grown accustomed to using the Internet, to the point where it appears hard to function without it. Daily tasks, including housework, require the Internet. Over time, it appears that technology is progressively becoming more focused on people. Technology's advancement will eliminate all forms of opaqueness between individuals, enterprises, and objects. Information technology is developing and expanding quickly these days, profoundly altering both our personal and professional lives in several ways [2]. Industrial revolutions have always occurred in tandem with the numerous advancements made by industry, a vital component of a nation's economy, throughout history. These advancements have fundamentally altered manufacturing methods. The Fourth Industrial Revolution forced businesses to develop new strategies and invest in cutting-edge technologies to optimize productivity while consuming the fewest resources possible. The advent of the Fourth Industrial Revolution in recent times has brought about a transformation in the industry's atmosphere by facilitating the interconnection of various sections of a factory's production processes and tools [3, 4].

Figure 13.1 shows a block diagram of the IoT. It is a rapidly expanding network of sensors spread across a vast array of items or objects. It is anticipated that the main IoT components—sensors, actuators, and communication networks—will lead to automation across all industries. The main goal is to create a smart environment that enables data collection and sharing among all entities. IoT aims to make life better for everyone by reducing complicated physical systems to virtual ones. According to the National Institutes of Health (NIH) Statement on Sharing Scientific Research and the Gartner hype cycle from 2016, the IoT is currently one of the hottest topics. According to research conducted by Cisco, there will only be approximately 50 billion Internet-connected gadgets by 2020—roughly 2.78% of the 1.8 trillion things which are technically capable of being connected. Additionally, it is projected that there will be six times as many Internet-connected devices as people in the world in the future [1].

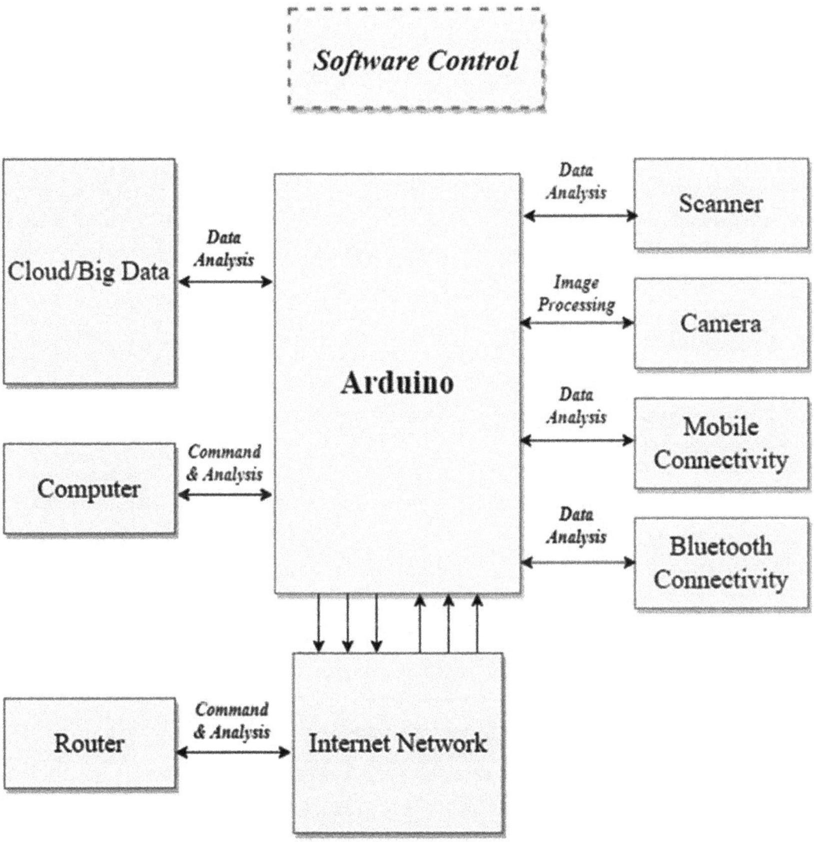

FIGURE 13.1 Internet of things (IoT).

The term "IoT" describes the complex interconnection of network connections, sensors, hardware, software, and everyday objects or devices. Both connected and wireless internet connections are possible, as are living and nonliving objects. Devices called sensors are embedded in or affixed to objects having the capacity to sense, read, and store data for later use in analysis. These items or equipment allow us to simplify, automate, track, or analyze our daily tasks more effectively. One such item is a basic smartphone that we carry. Sensors, sensor n/w, real-time localization, machine learning (ML), and other core components of the IoT have been used in the information technology industry for many years [5]. The term IoT can be termed from three different angles: the perspective of things or the electronic devices that will serve as the sensing objects; the outlook of the net or the perspective of interpretation, or the protocols in communication determined framework to which many objects are connected that handle the processing [6]. The IoT is regarded as a means of enabling universal computing across objects that have individual addressing schemes and the capacity to share data and communicate with one another. With the help of the IoT, people and things will be able to connect to anything at any time, anywhere, and via any network, path, service, or communication method [7] (Figure 13.2).

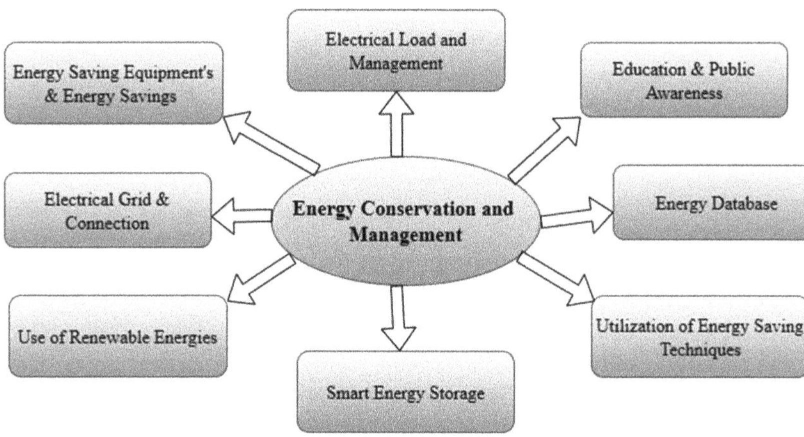

FIGURE 13.2 Energy conservation and management techniques.

Mathi et al. [7] highlighted the limitations of traditional farming methods, which often result in lower crop yields and increased human labor. To address these challenges, they introduced an IoT-based smart farming solution. This system focused on automating irrigation and pest detection using machine learning algorithms, particularly *K*-nearest neighbor and support vector machines. These algorithms utilized features extracted from plant leaves to accurately identify whether the plants are infected by pests. The proposed system not only monitors and analyzes agricultural fields but also achieves an impressive accuracy rate of 84% in disease detection, demonstrating its potential to improve crop productivity and reduce manual labor in farming. Zakaria et al. [16] presented an energy-monitoring device using ESP32 in an IoT system to track energy usage in a fused deposition modeling machine. The energy consumed was measured and categorized based on the machine's operational states, including idle, warm-up, printing, and cool-down. This innovative IoT system can simultaneously monitor various machines within a production system, serving as a foundational technology for multimachine energy monitoring. To increase food security and agricultural sustainability, Morchid et al. [9] studied applications of the IoT and sensor technology. This study focused on the prospective growth of the modern equipped agriculture market from 2021 to 2030. Furthermore, it elucidated the four tiers of IoT architecture tailored for smart agriculture, encompassing perception and actuation, networking, cloud, and applications. The review evaluated the current landscape of IoT and sensor technologies in agriculture, exploring their diverse applications, including irrigation monitoring, fertilizer management, disease detection, yield and quality monitoring, climate tracking, and fire prevention. The study also identified a variety of sensors capable of measuring crucial parameters in agriculture. Subsequently, it underscored the benefits of IoT in smart agriculture, such as enhanced efficiency, scalability, resource conservation, sustainability, agility, and product quality. Nevertheless, the study acknowledged the existing challenges that must be addressed to fully integrate IoT technology into agriculture, offering insights into potential future research directions. The research aims to provide valuable insights to readers and researchers seeking a comprehensive understanding of the current academic landscape in this field.

Mostofa and Islam [10] demonstrated an IoT-based system for the remote and real-life energy monitoring of solar photovoltaic facilities. They concluded that renewable energy sources, a reliable and widely accepted solution, offer an ideal alternative to fossil fuels for meeting our increasing energy needs. Solar energy, which can be converted into electricity using solar cells, is commonly deployed in remote locations such as rooftops, mountains, and deserts. To optimize the performance of these solar photovoltaic systems, effective monitoring is essential. Various monitoring solutions exist, including a basic offline data collection method, where data is stored locally for a specific period using a data collection device placed on the installation. However, this conventional approach falls short of providing real-time data. In contrast, harnessing IoT technology for real-time remote monitoring of solar photovoltaic installations offers a significant performance boost, with data stored in the cloud and accessible worldwide through an application. This system enables immediate insights into installation status, streamlining maintenance, and timely fault detection. Chiradeja et al. [18] proposed an innovative approach to developing public-lighting systems integrated with IoT for smart cities. LED lights, controlled by light intensity and motion sensors, regulate brightness to meet standard requirements. Smart lighting operates in manual, scheduled, and auto modes, each tested and compared for lighting and power quality. Beyond lighting, IoT features included air pollution detection, video surveillance, and flood-warning systems. The work presented a prototype with parameters optimized using DIALux software. Energy savings were achievable in auto mode, although it causes minor electrical grid harmonic distortion. IoT system precision relied on environmental factors and wireless network reliability. Shvets and Hanák [11] conducted a comprehensive review of the Internet of Things (IoT) and its impact on operations management, drawing insights from a wide range of literature. It primarily focused on analyzing a substantial collection of scholarly articles with the keywords IoT and operations management in the Scopus and Web of Science (WOS) databases. The analysis incorporated various methodologies, including bibliometric analysis, life cycle analysis, and text analysis to categorize IoT applications in operations management into five clusters: digitization, digital operations, monitoring system, tracking, and certification. This research also highlighted the identification of influential researchers, universities, and countries, contributing to a deeper understanding of the IoT's effects on operational management, and ultimately advancing the field of study.

Ding et al. [2] revealed the hidden potentials of the IoT through an integrated approach using agent-based modeling and system dynamics to assess sustainable supply-chain performance. The IoT offered novel prospects for optimizing eco-friendly and cost-effective supply chains. This study quantified such enhancements in a specific logistics case, employing system dynamics (SD) and agent-based modeling (AB) to compare IoT-backed and traditional logistical systems. Specifically, it evaluated the disparities in carbon emissions, costs, and market performance in a battery delivery chain within these two frameworks. The results revealed that IoT supply chains excel in reducing product storage and shipment fluctuations while boosting overall profitability by over 30%. Emissions from heating, lighting, and transportation per unit substantially decreased under various demand scenarios. Nevertheless, the advantages of IoT may be diminished due to increased heavy truck use. Despite a 10% lower market share, IoT products can match conventional ones with a value-added approach instead of pricing. Singh et al. [12] described the Internet of Things in the construction industry and facility management. They found that the Construction 4.0 era is witnessing rapid IoT evolution, playing a pivotal role in the digitalization process. They explored applications in construction, focusing on both the construction and operational phases of facility life cycles. It gathered and discussed examples, supplemented by relevant review papers, revealing IoT's diverse potential. The usage of IoT varies across different life cycle phases, with an emphasis on quality, safety, and health during construction, and smart concepts and energy management in the operational phase. Notably, both phases emphasize the interplay between Building Information Modeling and IoT, which were expected to grow in importance as these technologies become more prevalent in future construction projects.

13.1.1 SMART HOME

To comprehend the true significance of the IoT and how it will impact our lives in the next years, let us take a look at a basic smart house example. Let us imagine someone leaves for work and forgets to turn off the lights or air conditioning at his home. Thanks to smart home technology, he can accomplish this remotely from outside his home. Similarly, the air conditioner in his living room automatically adjusts the temperature when he enters his home in the evening after work based on information gathered from temperature sensors positioned all over his body [13] (Figure 13.3).

Additionally, the refrigerator uses data from sensors placed on his body to suggest foods like fruits or energy drinks to help him regain energy. The music system plays music that suits his mood, such as rock or soft music, and the lights change to suit his eyes or comfort level. This is what the adoption of IoT can bring about in a few years. This demonstrates how the IoT will undoubtedly enhance our lives. IoT is pervasive across many industries and offers a vast array of features [14]. Together with significant income growth, this will also provide a lot of job chances.

These industries include those focused on consumers, the environment, health, etc. Therefore, it can be concluded that IoT will see rapid expansion in the upcoming years, offering enormous incentives and chances for study, application, and employment. The IoT has integrated wireless sensor network technologies that facilitate the global interconnection of smart devices with advanced functionalities [15]. The key technology in the creation of intelligent homes is a wireless home automation network, which is made up of sensors and actuators that share resources and are interconnected to each other. A smart home is an IoT paradigm that integrates home automation. By connecting objects and devices in a home to the Internet, users can monitor and control them remotely [8].

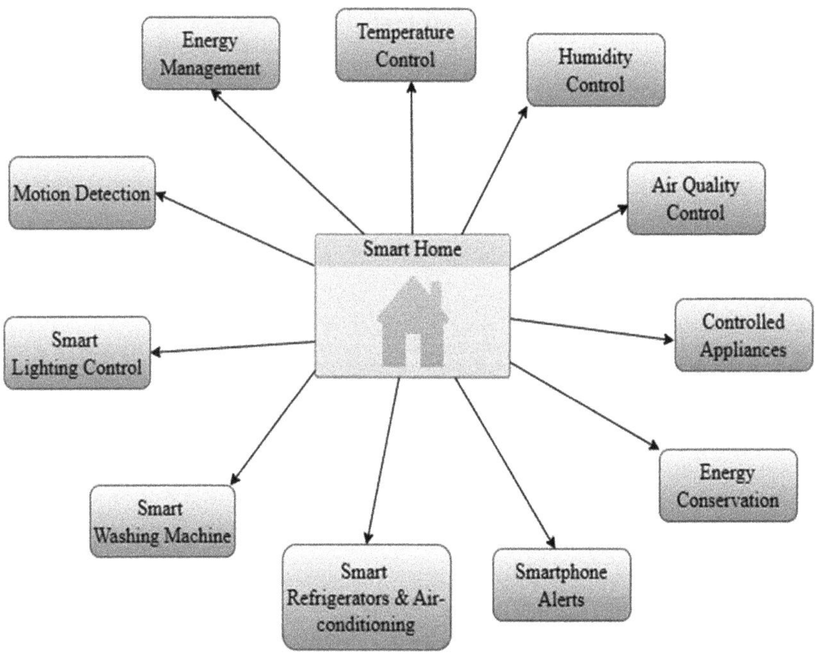

FIGURE 13.3 Smart house.

13.1.2 HISTORY OF SMART HOME

The first vacuum cleaner was developed in 1901 together with other appliances. The vacuum cleaner itself was established in 1907. After a while, more appliances such as electric dryers, freezers, and irons were created in 1966–1967. However, the ECHO IV was the first device to manage the temperature of the home as well as the on and off of appliances/devices for lighting control and security monitoring in the 1970s [16]. Half of England's population owned a video recorder in 1990, while the majority of people had color television in the 1980s. Aging technology to support senior citizens in 1991. In 1994, consumers started using rotary dryers more frequently in addition to multimedia computers, DVD players, cordless phones, and PlayStations.1998 offered new networks for communication, including integrated services digital network (ISDN) and the Internet, as well as accessories like web television and video phones [17]. With the introduction of various low-cost technology in the year 2000, the smart house trend started to gain traction. The first wearable smart health system was released in 2005. 2009 saw the integration of the cloud into the smart home system. 2015 saw the adoption of voice as a controller for smart homes [13, 14].

13.2 ENERGY CONSERVATION AND MONITORING USING INTERNET OF THINGS

Figure 13.4 shows the block diagram of the use of IoT for energy conservation and monitoring. It consists of an Arduino Uno board, a camera, an Image processing unit, a display unit, a proximity sensor, a temperature sensor, a speed monitoring unit, a gas leakage detector, an Amazon and Alexa Integration unit, an energy and power management unit, smart diagnosis and IoT platform. These devices are used to process, control, and display the results to monitor and manage the energy and other parameters.

13.2.1 ARDUINO UNO WITH PROXIMITY SENSOR

The proximity sensor is used to detect objects physically contacting them. It is a noncontact detection device that is used to compare the sensor's outputs, limits, etc. Figure 13.5 shows a circuit diagram of Arduino with a proximity sensor for monitoring the energy parameters.

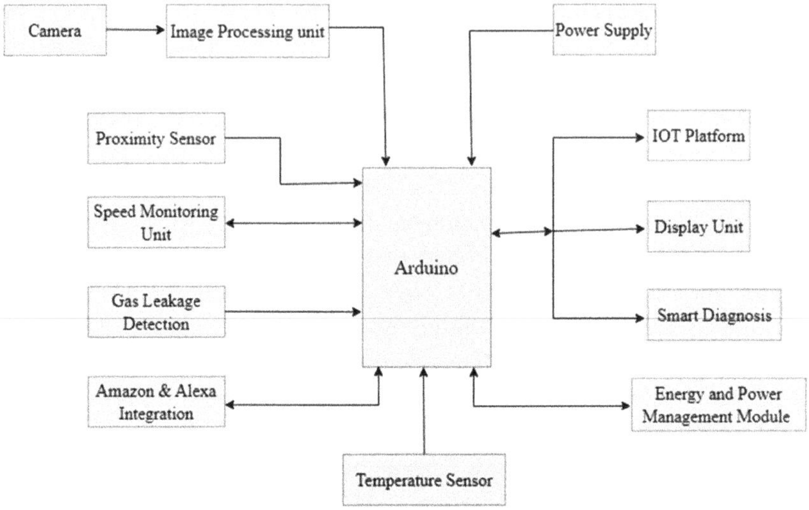

FIGURE 13.4 Use of IoT for energy conservation and monitoring.

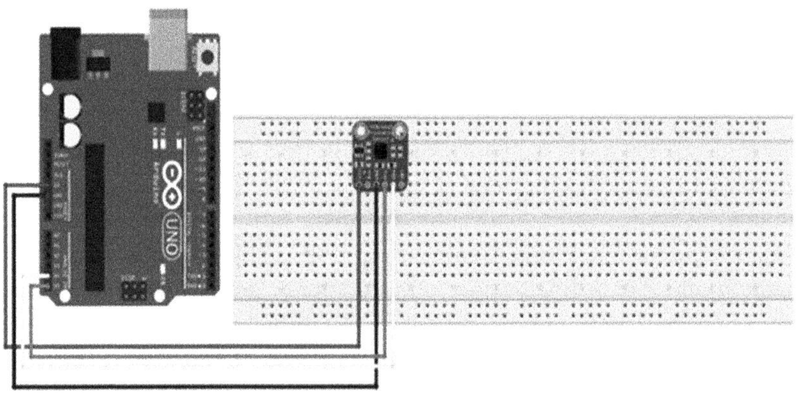

FIGURE 13.5 Circuit diagram of Arduino with proximity sensor.

13.2.2 ARDUINO UNO TEMPERATURE SENSOR WITH DISPLAY DEVICE LCD

Display units are devices like LCD, LED, etc. that are used to display output and to analyze the parameter differently with or without wire. Figure 13.6 shows the circuit diagram of the Arduino board with a temperature sensor. The temperature sensor is a device that is used to detect temperature in electrical analog signals. This analog signal is converted into a digital signal to monitor and for further processing.

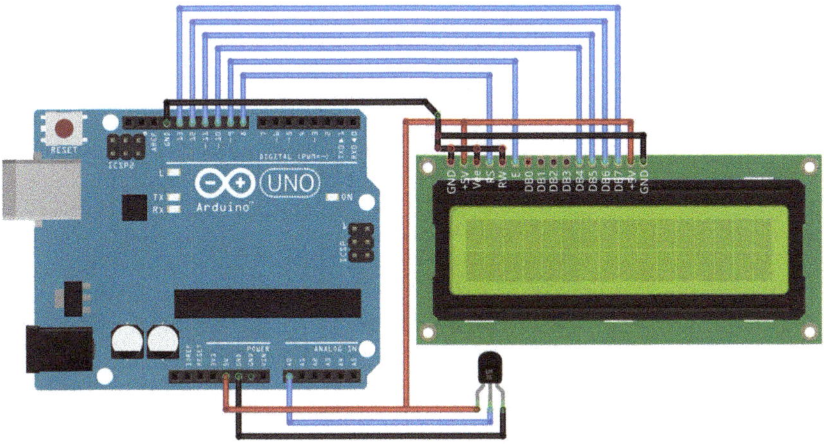

FIGURE 13.6 Circuit diagram of Arduino with temperature sensor.

13.2.3 ARDUINO UNO WITH GAS LEAKAGE DETECTOR

A gas leakage sensor is a device that is used to detect and measure the concentration of different gases (Figure 13.7). It is also notified of the different levels of the gas in terms of light indicators and comparative analysis of the different gases.

13.2.4 IoT PLATFORM

The IoT platform consists of many devices like the Internet, processing devices, and other peripherals used in this platform. It is used to monitor, process, compare, and display various parameters to monitor with or without wire.

13.2.5 SPEED MONITORING UNIT

This unit is used to detect and compare the speed system. It is also responsible for processing and analyzing the data for monitoring and comparing.

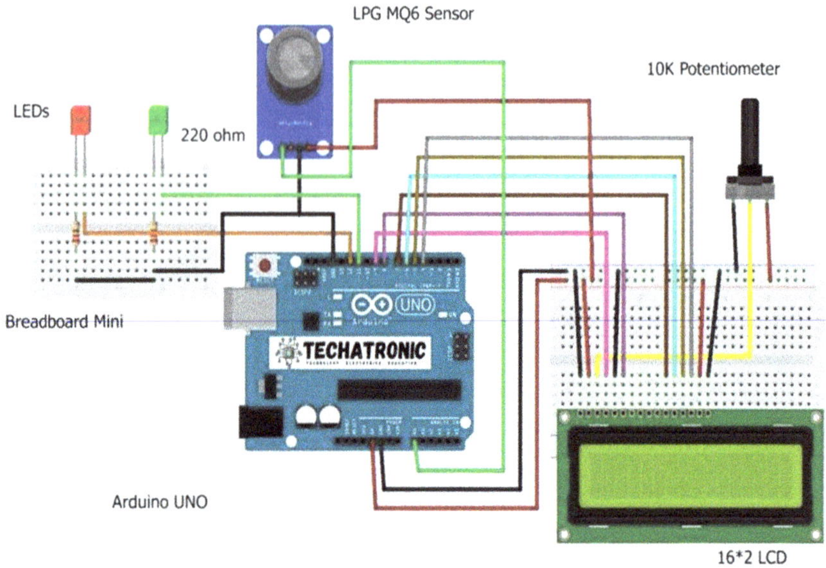

FIGURE 13.7 Circuit diagram of Arduino with gas leakage sensor.

13.2.6 SMART DIAGNOSIS

Manufacturers of smart refrigerators can enhance the capabilities of their sensors through firmware integration. By altering the programming of the sensors, manufacturers can expand the sensors' broad applications and enhance their capacity to recognize specific elements. Manufacturers can use program sensors to detect particular compounds, such as ethylene, that might accelerate food deterioration. The sensors' ability to counsel customers on whether to eat or dump the affected items is another special feature. Sensors can measure and monitor the levels of humidity and temperature more accurately, thanks to firmware integration. The duration of shelf life and nutritional value of perishable foods are extended.

13.2.7 AMAZON AND ALEXA INTEGRATION

Alexa and other voice assistants, as well as online retailers like Amazon, can be interconnected with smart refrigerators. They provide the ability to order groceries directly from their refrigerator, create shopping lists automatically, and even seek up recipes utilizing ingredients that are in the refrigerator attributable to this connectivity. It improves the efficiency and convenience of food preservation with shopping.

13.2.8 ENERGY AND POWER MANAGEMENT

Modern smart refrigerators are equipped with a cutout that shuts down the compressor once the pressure inside the line reaches to peak. It helps to save energy. One can refer to a trendy refrigerator equipped with sensors, connection, and intelligent functions as a smart fridge (Figure 13.8). It gives proprietors a more useful and effective way to manage their food stores. Several internal technologies allow smart refrigerators to communicate with other intelligent appliances, humans, and even the Internet, which sets them apart from regular refrigerators.

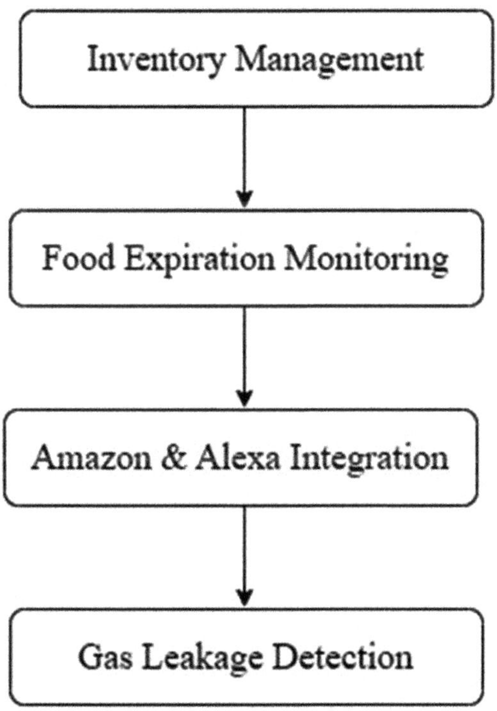

FIGURE 13.8 Energy conservation and monitoring in refrigeration system using IoT inventory management.

End users may simply manage and remotely monitor the storage space of their interconnected refrigerator by using a dedicated smartphone app. Customers can assess their present possessions by using the app, which displays the fridge's inventory. Original equipment manufacturers (OEMs) can market these features as a way to monitor inventory and receive warnings when specific products run low.

13.2.8.1 FOOD EXPIRATION MONITORING

When food is about to expire, smart refrigerators can be controlled by using different sensors. Users are instantly notified when products are about to expire through timely alerts. This feature ensures that consumers eat healthier food and promotes better inventory control by reducing food waste.

13.2.8.2 NUTRITIOUS FOOD UNDERSTANDING THROUGH IMAGE RECOGNITION

Discover how smart refrigerators use cutting-edge technology for purposes beyond storing. These refrigerators can determine the nutritional value of the food they hold because of advanced sensors and digital image recognition technology. This feature helps customers make better eating choices by providing them with relevant information about their dietary habits.

13.2.9 USE OF TEMPERATURE SENSORS FOR TEMPERATURE OPTIMIZATION

Advanced temperature control can be used to achieve excellent cooling, and modern technology-based refrigerators are employed (Figure 13.9). They use sensors to check the temperature inside and make necessary adjustments. With this method, a lower chance of food spoiling can be achieved for longer periods at the appropriate temperature in the refrigerator.

13.2.10 COMPRESSOR SPEED MONITORING

A vital component of a refrigerator's cooling system is the compressor. Smart refrigerators use technology that modifies the speed of the compressor following cooling demands. The refrigerator's speed adjustment feature adjusts its operation based on usage patterns and contents to help reduce energy use and maintain reliable cooling.

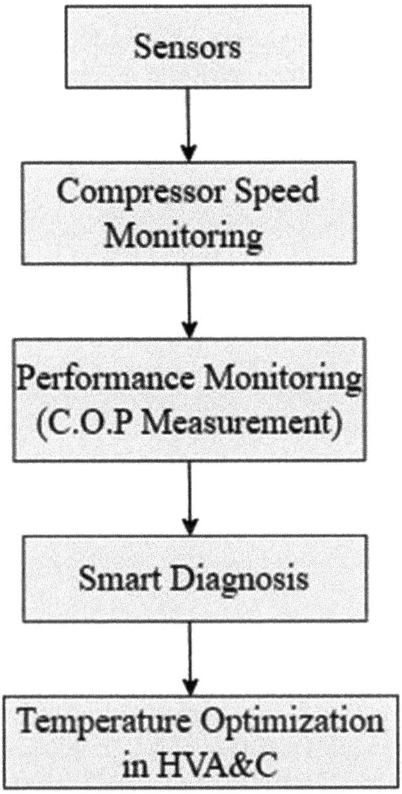

FIGURE 13.9 Energy conservation and monitoring in HVA&C using IoT.

13.2.11 PERFORMANCE MONITORING

Smart refrigerators come equipped with sensors that continuously monitor several performance-related aspects. These sensors can identify variations in temperature, humidity, and several other elements that affect how well the refrigerator functions. By tracking real-time performance and making necessary adjustments, it may be able to detect any problems or abnormalities.

13.3 CONCLUSION

The IoT has the potential to enhance energy management and conservation through the management of electrical loads, energy storage, linkages between electrical network systems, energy saving in buildings and industries, as well as public awareness and education. It is recommended that renewable energy be integrated into current energy systems as part of an overall energy policy, with the adoption of such new technologies expected to contribute between 15% and 20% over the next few decades.

KEYWORDS

- **energy monitoring**
- **energy conservation**
- **IoT**

REFERENCES

1. Constance, E. The Internet of Things: Preparing for the revolution. *Journal of Cyber Policy* **2017**, *2*(2), 152–154.
2. Ding, S., Ward, H., Cucurachi, S., & Tukker, A. Revealing the hidden potentials of Internet of Things (IoT)—An integrated approach using agent-based modelling and system dynamics to assess sustainable supply chain performance. *Journal of Cleaner Production* **2023**, *421*, 138558. https://doi.org/10.1016/j.jclepro.2023.138558
3. Eltamaly, A. M., Alotaibi, M. A., Alolah, A. I., & Ahmed, M. A. IoT-based hybrid renewable energy system for smart campus. *Sustainability* **2021**, *13*(15), 1–18.
4. Tripathi, S. L., Ganguli, S., Kumar, A., & Magradze, T. eds. *Intelligent Green Technologies for Sustainable Smart Cities*. John Wiley & Sons, **2022**.
5. Hu, J., van der Vlist, B., Niezen, G., Willemsen, W., Willems, D., & Feijs, L. Designing the Internet of Things for learning environmentally responsible behaviour. *Interactive Learning Environments* **2013**, *21*(2), 211–226. https://doi.org/10.1080/10494820.2012.705858
6. Yohesh, M., Andro, E. A., Aswin, S. R., Kamalesh, S., Parthasaradi, V.. Smart home automation system based on Li-Fi technology. *International Journal of Innovative Research in Engineering* **2023**, *4*(3), 342–345. https://doi.org/10.59256/ijire.2023040397
7. Mathi, S., Akshaya, R., & Sreejith, K. An internet of things-based efficient solution for smart farming. *Procedia Computer Science* **2022**, *218*, 2806–2819.

8. Mendez Mena, D., Papapanagiotou, I., & Yang, B. Internet of things: Survey on security. *Information Security Journal* **2018**, *27*(3), 162–182.
9. Morchid, A., El Alami, R., Raezah, A. A., & Sabbar, Y. Applications of internet of things (IoT) and sensors technology to increase food security and agricultural Sustainability: Benefits and challenges. *Ain Shams Engineering Journal* **2023**, *15*(3), 102509. https://doi.org/10.1016/j.asej.2023.102509
10. Mostofa, K. Z., & Islam, M. A. Creation of an Internet of Things (IoT) system for the live and remote monitoring of solar photovoltaic facilities. *Energy Reports* **2023**, *9*, 422–427.
11. Shvets, Y., & Hanák, T. Use of the internet of things in the construction industry and facility management: Usage examples overview. *Procedia Computer Science* **2023**, *219*, 1670–1677.
12. Singh, N., Raza, M., Paranthaman, V. V., Awais, M., & Khalid, M. Internet of things and cloud computing. In: *Digital Health. INC* **2021** https://doi.org/10.1016/B978-0-12-818914-6.00013-2
13. Sinha, S., Raj, T. V., Kumar, S. Student & Minor Project, *Smart Home Automation Using IoT with Cloud Computing. February*, **2023**, https://www.researchgate.net/publication/368838967
14. Stolojescu-Crisan, C., Crisan, C., & Butunoi, B. P. An IoT-based smart home automation system. *Sensors* **2021,** *21*(11), 1–23. https://doi.org/10.3390/s21113784
15. Taiwo, O. Internet of things-based intelligent smart home control system. *Security and Communication Networks*. **2021**, *2021*(7), 1–17
16. Zakaria, S., Mativenga, P., & Ariff, E. A. R. E. An investigation of energy consumption in fused deposition modelling using ESP32 IoT monitoring system. *Procedia CIRP* **2023** *116*, 263–268. https://doi.org/10.1016/j.procir.2023.02.045
17. Zhang, H., Li, F., Wang, J., Wang, Z., Sanín, C., & Szczerbicki, E. Experience-oriented intelligence for internet of things. *Cybernetics and Systems* **2017** *48*(3), 162–181. https://doi.org/10.1080/01969722.2016.1276771
18. Chiradeja, P., & Yoomak, S. Development of public lighting system with smart lighting control systems and internet of thing (IoT) technologies for smart city. *Energy Reports* **2023**, 10(August), 33553372. https://doi.org/10.1016/j.egyr.2023.10.027

Index

R

S

U

V

W

www.ingramcontent.com/pod-product-compliance
Ingram Content Group UK Ltd.
Pitfield, Milton Keynes, MK11 3LW, UK
UKHW022159190926
475702UK00003B/102
* 9 7 8 1 7 7 9 6 4 3 5 8 2 *